Studies in Computational Intelligence

Volume 985

The series "Studies in Computational Intelligence" (SCI) publishes new developments and advances in the various areas of computational intelligence—quickly and with a high quality. The intent is to cover the theory, applications, and design methods of computational intelligence, as embedded in the fields of engineering, computer science, physics and life sciences, as well as the methodologies behind them. The series contains monographs, lecture notes and edited volumes in computational intelligence spanning the areas of neural networks, connectionist systems, genetic algorithms, evolutionary computation, artificial intelligence, cellular automata, self-organizing systems, soft computing, fuzzy systems, and hybrid intelligent systems. Of particular value to both the contributors and the readership are the short publication timeframe and the world-wide distribution, which enable both wide and rapid dissemination of research output.

Indexed by SCOPUS, DBLP, WTI Frankfurt eG, zbMATH, SCImago.

All books published in the series are submitted for consideration in Web of Science.

More information about this series at http://www.springer.com/series/7092

Roger Lee
Editor

Computer and Information Science 2021—Summer

Springer

Editor
Roger Lee
Software Engineering and Information
Technology Institute
Central Michigan University
Mount Pleasant, MI, USA

ISSN 1860-949X ISSN 1860-9503 (electronic)
Studies in Computational Intelligence
ISBN 978-3-030-79473-6 ISBN 978-3-030-79474-3 (eBook)
https://doi.org/10.1007/978-3-030-79474-3

This Springer imprint is published by the registered company Springer Nature Switzerland AG
The registered company address is: Gewerbestrasse 11, 6330 Cham, Switzerland

Foreword

The purpose of the 20th IEEE/ACIS International Summer Semi-Virtual Conference on Computer and Information Science (ICIS 2021) held on June 23–25, 2021, in Shanghai, China, was to bring together researchers, scientists, engineers, industry practitioners, and students to discuss, encourage, and exchange new ideas, research results, and experiences on all aspects of computer and information science, and to discuss the practical challenges encountered along the way and the solutions adopted to solve them. The conference organizers have selected the best 13 papers from those papers accepted for presentation at the conference in order to publish them in this volume. The papers were chosen based on review scores submitted by members of the program committee and underwent further rigorous rounds of review.

In the chapter "The Effect of Online Investor Sentiment on Stock Movements: An LSTM Approach," Han Wang, Liang Xue, Wencai Du, Fengling Wang, Peng-sheng Li, Lijin Chen, and Huawei Ma analyzing stock trends based on the sentiment of social media provide a novel direction for investors to analyze the stock market.

In the chapter "A Framework and Decision Algorithm to Determine the Best Feature Extraction Technique for Supporting Machine Learning-Based Hate Speech Detection," Chun-Kit Ngan and Kashyap Bhuva develop and implement a framework and a decision algorithm to determine the best feature extraction technique (FET) for supporting machine learning-based hate speech detection.

In the chapter "Sentiment Analysis of Stock Market Investors and Its Correlation with Stock Price Using Maximum Entropy," Liang Xue, Han Wang, Fengling Wang, and Huawen Ma use a Python web crawler to extract and preprocess the crawled data, into a classifier model using maximum entropy which is built to classify stock-related comments into three sentiment labels for analysis.

In the chapter "Intrusion Detection for Modern DDos Attacks Classification Based on Convolutional Neural Networks," Wenwen Chen, Haiyan Zhang, Xiaoshu Zhou, and Yangjie Weng propose an intrusion detection measure to classify modern DDoS attacks using convolutional neural network (CNN), a deep learning algorithm.

In the chapter "KnowGraph-PM: A Knowledge Graph Based Pricing Model for Semiconductor Supply Chains," Nour Ramzy, Sören Auer, Javad Chamanara, and Hans Ehm evaluate a knowledge graph-based dynamic pricing model approach by calculating the revenue generated after applying the pricing algorithm.

They demonstrate that semantic data integration enables customer-tailored revenue management.

In the chapter "A Study on the Recognition of Hangeul Through Transitional Learning in Handwritten Application," Jae Hyuk Heo, Sang Wook Lee, Hee Won Lee, and Gwang Yong Gim study the improvement method for optical character recognition (OCR) of handwritten general educational development (GED) application.

In the chapter "Study on Partial Image Detection for Drawing—Focus on Unstructured Images Included in the Main Image," Ji Won Lee, Jae Ho Lee, Doh Yeon Kim, and Gwang Yong Gim present a study that aims to learn images by applying deep learning so that they can be applied to picture projection tests among unusual fields and to detect characteristic images that can be used for psychological interpretation.

In the chapter "Factors Affecting the Intention to Use Artificial Intelligence-Based Recruitment System: A Structural Equation Modeling (SEM) Approach," Jung Hee Lee, Ju Hyung Kim, Yong Hwan Kim, Yong Min Song, and Gwang Yong Gim analyze the factors affecting the intention of use of AI-based recruitment system by utilizing TOE and TAM. As a result, it was shown that the reliability, security, suitability, new technology, partiality, readiness, and legal and policy environment of the TOE affected the intention of using the system.

In the chapter "A Comparative Study of Vectorization Approaches for Detecting Inconsistent Method Names," Tomoya Minehisa, Hirohisa Aman, Tomoyuki Yokogawa, and Minoru Kawahara present a comparative study of vectorization approaches for detecting inconsistent method names which focuses on such a computational cost and proposes to replace it with another lightweight vectorization approach.

In the chapter "Heart Sound Segmentation Based on a Joint HSMM Method," Yao Zhang, Zeyu Ma, Xin Zhou, Xianhong Li, Ying Liu, Mingang Chen, Xin Sun, Xuying Wang, Jingtao Wang, Lizhi Cai, and Kun Sun propose a novel joint HSMM method that combines the CNN and probabilistic models (HSMM) for heart sound segmentation.

In the chapter "A Novel Authentication System for Artwork Based on Blockchain," Seung Wook Jung provides a novel solution for blockchain air gap, between the blockchain listing and physical artworks that addresses the problem efficiently but not expensively.

In the chapter "Image Steganography Using GANs," Ketan Ramaneti, Pranavi Kakani, Chaitanya Krishna, and Sujatha Rajkumar propose a system that eliminates risk by hiding secret information inside stego images generated using generative adversarial networks (GANs) and then safely reproducing it using an extractor model.

In the chapter "Coverage-Guided Fairness Testing," Daniel Perez Morales, Takashi Kitamura, and Shingo Takada propose a coverage-guided fairness testing (CGFT). CGFT leverages combinatorial testing to generate an evenly distributed test suite.

It is our sincere hope that this volume provides stimulation and inspiration, and that it will be used as a foundation for works to come.

June 2021

Jiayu Gong
Shanghai Key Laboratory of Computer
Software Testing and Evaluating
Shanghai, China

Contents

Contributors

Hirohisa Aman Center for Information Technology, Ehime University, Matsuyama, Ehime, Japan

Sören Auer TIB Leibniz Information Centre for Science and Technology and L3S Research Center, Leibniz University of Hannover, Hannover, Germany

Kashyap Bhuva Data Science Program, Worcester Polytechnic Institute, Worcester, MA, USA

Lizhi Cai Shanghai Key Laboratory of Computer Software Testing & Evaluating, Shanghai Development Center of Computer Software Technology, Shanghai, China

Javad Chamanara TIB Leibniz Information Centre for Science and Technology and L3S Research Center, Leibniz University of Hannover, Hannover, Germany

Mingang Chen Shanghai Key Laboratory of Computer Software Testing & Evaluating, Shanghai Development Center of Computer Software Technology, Shanghai, China

Lijin Chen Faculty of International Tourism and Management, City University of Macau, Macau, China;
Tianmu Cultural Tourism Construction Co., Ltd., Zhuhai, China

Wenwen Chen Department of Computer Technology, Beijing Institute of Technology, Tangjiawan, Zhuhai, Guangdong, China

Wencai Du Institute of Data Science, City University of Macau, Macau, China

Hans Ehm Infineon Technologies AG, Neubiberg, Germany

Gwang Yong Gim Department of Business Administration, Soongsil University, Seoul, Korea

Jae Hyuk Heo Department of IT Policy and Management, Graduate School, Soongsil University, Seoul, Korea

Seung Wook Jung Department of Cyber Security, Konyang University, Nonsan, Korea

Pranavi Kakani School of Electronics Engineering, Vellore Institute of Technology, Vellore, India

Minoru Kawahara Center for Information Technology, Ehime University, Matsuyama, Ehime, Japan

Doh Yeon Kim Department of IT Policy and Management, Graduate School, Soongsil University, Seoul, Korea

Ju Hyung Kim Department of IT Policy and Management, Graduate School of Soongsil University, Seoul, Korea

Yong Hwan Kim Department of IT Policy and Management, Graduate School of Soongsil University, Seoul, Korea

Takashi Kitamura National Institute of Advanced Industrial Science (AIST), Tokyo, Japan

Chaitanya Krishna School of Electronics Engineering, Vellore Institute of Technology, Vellore, India

Hee Won Lee Department of IT Policy and Management, Graduate School, Soongsil University, Seoul, Korea

Ji Won Lee Department of IT Policy and Management, Graduate School, Soongsil University, Seoul, Korea

Jae Ho Lee J and Lee Co., Koung-gi, Korea

Jung Hee Lee Department of Business Administration, Graduate School of Soongsil University, Seoul, Korea

Sang Wook Lee Department of IT Policy and Management, Graduate School, Soongsil University, Seoul, Korea

Pengsheng Li Institute of Data Science, City University of Macau, Macau, China; School of Information Technology, Beijing Institute of Technology, Zhuhai, China

Xianhong Li Enterprise Research Institute, Hangzhou Ewell Technology Company, Hangzhou, China

Ying Liu Department of Pediatric Cardiology, Xinhua Hospital affiliated to Shanghai Jiaotong University School of Medicine, Shanghai, China

Zeyu Ma Shanghai Key Laboratory of Computer Software Testing & Evaluating, Shanghai Development Center of Computer Software Technology, Shanghai, China

Huawei Ma Institute of Data Science, City University of Macau, Macau, China; School of Information Technology, Beijing Institute of Technology, Zhuhai, China

Huawen Ma Institute of Data Science, City University of Macau, Macau, China; School of Information Technology, Beijing Institute of Technology, Zhuhai, China

Tomoya Minehisa Graduate School of Science and Engineering, Ehime University, Matsuyama, Ehime, Japan

Daniel Perez Morales Keio University, Yokohama, Japan

Chun-Kit Ngan Data Science Program, Worcester Polytechnic Institute, Worcester, MA, USA

Sujatha Rajkumar School of Electronics Engineering, Vellore Institute of Technology, Vellore, India

Ketan Ramaneti School of Electronics Engineering, Vellore Institute of Technology, Vellore, India

Nour Ramzy Leibniz University of Hannover, Hannover, Germany

Yong Min Song Department of Business Administration, Graduate School of Soongsil University, Seoul, Korea

Kun Sun Department of Pediatric Cardiology, Xinhua Hospital affiliated to Shanghai Jiaotong University School of Medicine, Shanghai, China

Xin Sun Clinical Research Unit, Xinhua Hospital affiliated to Shanghai Jiaotong University School of Medicine, Shanghai, China

Shingo Takada Keio University, Yokohama, Japan

Fengling Wang Institute of Data Science, City University of Macau, Macau, China; Guangdong University of Education, Guangzhou, Guangdong, China

Han Wang Institute of Data Science, City University of Macau, Macau, China; Zhuhai Institute of Advanced Technology, Chinese Academy of Sciences, Zhuhai, China

Jingtao Wang Xunyin Intelligent Technology (Shanghai) Co., Ltd., Shanghai, China

Xuying Wang Enterprise Research Institute, Hangzhou Ewell Technology Company, Hangzhou, China

Yangjie Weng Zhuhai Central Sub-Branch, The People's Bank of China, Zhuhai, Guangdong, China

Liang Xue Institute of Data Science, City University of Macau, Macau, China; Guangdong University of Education, Guangzhou, Guangdong, China

Tomoyuki Yokogawa Faculty of Computer Science and Systems Engineering, Okayama Prefectural University, Soja, Okayama, Japan

Haiyan Zhang Department of Computer Technology, Beijing Institute of Technology, Tangjiawan, Zhuhai, Guangdong, China

Yao Zhang Enterprise Research Institute, Hangzhou Ewell Technology Company, Hangzhou, China

Xiaoshu Zhou Department of Computer Technology, Beijing Institute of Technology, Tangjiawan, Zhuhai, Guangdong, China

Xin Zhou Clinical Research Unit, Xinhua Hospital affiliated to Shanghai Jiaotong University School of Medicine, Shanghai, China

The Effect of Online Investor Sentiment on Stock Movements: An LSTM Approach

Han Wang, Liang Xue, Wencai Du, Fengling Wang, Pengsheng Li, Lijin Chen, and Huawei Ma

Abstract Analyzing stock trends based on the sentiment of social media provides a novel direction for investors to analyze the stock market. Behavioral financial theory and social psychology indicate that irrational behaviors related to financial decisions could result in stock fluctuations. Taking representative 20 stocks on Shanghai Stock Exchange as an example, user generated contents from January 31, 2017 to January 31, 2019 are obtained from Sina and Fortune.com. TF-IDF and TextRank algorithms are applied to extract keywords, based on which 2000-word-level financial sentiment lexicon is generated. In addition, the LSTM model is built and 23,152 comments were analyzed based on the lexicon. Eventually, relationships between sentiment scores and the trend of stock fluctuation are explored by applying the correlation

H. Wang · L. Xue · W. Du · F. Wang · P. Li · H. Ma (✉)
Institute of Data Science, City University of Macau, Macau, China
e-mail: mkmhw@bitzh.edu.cn

H. Wang
e-mail: wanghan@ziat.ac.cn

W. Du
e-mail: georgedu@cityu.mo

F. Wang
e-mail: wf1232983@163.com

P. Li
e-mail: 12257@bitzh.edu.cn

H. Wang
Zhuhai Institute of Advanced Technology, Chinese Academy of Sciences, Zhuhai, China

L. Xue · F. Wang
Guangdong University of Education, Guangzhou, Guangdong, China

P. Li · H. Ma
School of Information Technology, Beijing Institute of Technology, Zhuhai, China

L. Chen
Faculty of International Tourism and Management, City University of Macau, Macau, China
e-mail: t16091100230@cityu.mo

Tianmu Cultural Tourism Construction Co., Ltd., Zhuhai, China

R. Lee (ed.), *Computer and Information Science 2021—Summer*,
Studies in Computational Intelligence 985,
https://doi.org/10.1007/978-3-030-79474-3_1

coefficient parameter and Apriori algorithm. Results show that LSTM has a great advantage in sentiment analysis, which presents a higher accuracy (99.87%) than the sentiment lexicon-based method (94.57%). Taking the delay impact of stockholders' sentiments on the stock trend into account, this research discusses the correlation between current investor sentiments and stock markets in the next few days. The paper finds that current emotional tendency has a deeper influence on the stock trend at the third day afterwards. Thus, this study extends financial sentiment lexicons, explores applications of LSTM machine learning in financial fields, and discusses the influence of investor sentiments on the stock market based on social media platforms. Processes of Web crawling, keyword extraction, sentiment analysis, correlation analysis and result visualization are coded in Python programming language, code packages are contributed through the Github website.

1 Introduction

The stock market plays an irreplaceable role in financial fields with its high return on investment. Social media platforms such as Sina and Easymoney.com stock forums, which contain a large amount of knowledge to be explored. Sentiment analysis based on machine learning and sentiment lexicons (SentiCon) plays a significant role for decision-making supporting. The behavioral finance theory and social psychology indicate that irrational behaviors related to financial decision-making are able to affect the stock market at a certain degree. Taking 20 stocks of Shanghai Stock Exchange from 2017 to 2019 as a study case, this study explores and compares the lexicon-based and LSTM sentiment analysis methods on user generated contents (UGC). Coding in Python programming language, this study collects data with web crawlers, builds a finance lexicon by applying keyword extraction and word cloud technologies. The effect of online sentiment on stock movements is discussed with LSTM sentiment results. Five contributions are concluded. (1) The stock financial corpus is expanded; (2) A new financial sentiment lexicon is constructed; (3) Sentiment analysis models based on the sentiment lexicon and LSTM are established and compared; (4) The relationship between stockholders' sentiment and stock trends is explored, providing investors with decision-making supports; (5) Core algorithms are uploaded to Github (https://github.com/luckanny111), hopping to be helpful for future researches.

2 Related Works

2.1 Researches on Traditional Stock Decision Analysis

Since the Dutch East India Company issued stocks to the public the first time, stocks have become more and more popular due to their high return on investments [1, 2]. Forecasting stock trends becomes the focus of attention of companies and individual

Table 1 Review of applications of sentiment analysis based on social media platforms

Methods	Surveys	Disadvantages
EMH, Efficient market hypothesis	Vachhani et al. [3], Rossi and Gunardi [4], Kumar and Jawa [5], and Shah et al. [6]	Make forecasts based on a relatively small amount of public information, but with a weak forecasting ability
RW, Random walk	Vachhani et al. [3], Shah et al. [6], Dash [7], Nasr et al. [8], Shaik and Maheswaran [9]	It is impossible to predict for stocks with greater volatility. However, the timeliness and stability are poor, real-time forecasts are difficult to achieve

stock owners [2]. However, it is always a challenge for stock predicting, especially for individuals. Conventional forecasting methods are divided into two aspects, Efficient Market Hypothesis (EMH) [3–6] and the Random Walk (RW, Random Walk) theory [3, 6–9]. The EMH theory states that asset prices reflect all available information for stock forecasts, which could be explained that there can be no better performance than the stock market for risk adjusting, as for stocks should only respond to new information. There aspects are concluded. First, weak EWH, it only considers the past transaction information. Second, semi-strong EMH, it uses all unbiased public information. Third, strong EMH, all information uses private. Random walk theory points out that stock market prices do not depend on the past data. Due to fluctuations, stock price prediction is seemly impossible [3]. The existence of price fluctuations in the stock market, undiscovered serial correlations between stock prices, basic events and economic effects are huge motivations for adopting better forecasting models. This precise prediction can be the basis for traders and investors, showing possibilities of stock market volatility [4]. In addition, due to the frequent data generation, real-time stock forecasting becomes one of the main challenges. With the latest development of cloud computing architecture [5–8], machine learning models are able to deployed on this high-performance cloud, which enhances the application of machine learning in stock forecasting (Table 1).

2.2 Researches on the Prediction of Stock Trend Based on Social Media and Sentiment Analysis

Sentiments are verified a significant cause for short-term market volatility, which can lead to discrepancies between the stock price and the true value of the company's shares. However, as the company's fundamentals eventually lead to the convergence of the share value and market price, weighing opportunity methods are still an vital strategy. Sentiment analysis based on various data sources is responsible to mining insights into the stock market's reaction to various news in the short and medium term.

Therefore, this novel method contributes a lot for investors by delivering decision making supports [6].

Sentiment analysis has been widely utilized in mining product reviews [10]. Some researchers have tried to analyze text data to improve stock market forecasts. There are two main sources of text data, which always be considered for this task, economic data and UGC in social media platforms [11–13]. Regarding to product and restaurant reviews, it is essential to extract topics and determine the sentiment polarities. Some preliminary studies decide the sentiment polarity by judge the adjectives nearby nouns [14]. Other prominent viewpoint mining models pay more attention on parsing sentences, extracting their grammar trees and finding the co-referential relationship of nouns [15, 16]. Another method of opinion mining is argument-based opinion mining, which uses argument theory to evaluate information. The framework is based on the abstract bipolar argumentation theory, which can extract relevant documents from text documents. Kalyanaraman et al. [17] combines argumentation theory with natural language processing methods, finding the most controversial arguments in the online shopping framework. Their introduced model combines argumentation and aspect-based views Dig [17]. Cakra and Trisedya [18] emphasize wide applications of sentiment analysis in academia and industries, their studies focus on understanding and mining insights under specific scenarios. Their paper suggests that bag-of-words models are the major method for emotion mining tasks. This research highlights the importance of integrating semantic knowledge in addition to machine learning methods. The author also suggests that the next-generation model will include common sense knowledge data and brain-inspired reasoning methods [18]. Gao et al. [19] introduces a Target-dependent sentiment classification method based on Bidirectional Encoder Representations (BERT), which presents a great advantage in aspect-based sentiment analysis tasks. According to the survey on the application of text mining in the financial field [20], conventional methods (such as decision trees, SVM and regression analysis) have been applied in 70% of the previous studies. According to another survey, in order to predict the market, most studies apply holistic methods (either feature level or decision level) to make predictions robust [21]. In recent years, with the emergence of deep neural networks, the application in finance research has increased dramatically. Recurrent Neural Networks (RNN) and Deep Belief Network (DBN) is employed for market prediction, results shows that machine learning methods could reduce the binary classification error rate from the baseline to 40.05%, and the error rate reached 47.30% [21]. Table 2 summarizes the literature research on sentiment analysis methods.

3 Methods

This paper utilizes Python to crawl for the k-line information of 20 stocks on the Shanghai Stock Exchange, as well as UGC of each stock in Sina and Easymoney.com stock forums from January 31, 2017 to January 31, 2019. Applying TF-IDF [22] and textRank [23] algorithms to extract keywords, a 2000-word-level investor emotional

Table 2 Review of applications of sentiment analysis to the stock price forecasting

Surveys	Data set	Methods	Unit	Standard	Results
Schumaker and Chen [12, 13]	News articles, S&P 500	Bag of words/noun phrases/noun entities and SVM	Daily	Returns, DA	2.57% (noun phrases)
Bollen et al. [14]	DJIA, Twitter data	Mood indicators and SOFNN	Daily	Accuracy	87.14%
Lee et al. [16]	8-K reports, stock prices, volatility	Ngram and random forest	Daily and long term	Accuracy	90%
Kalyanaraman et al. [17]	News articles (Bing API)	Lexicon approach and linear regression	Daily	Accuracy	81.82%
Pagolu et al. [18]	MSFT price, Twitter data	Ngram, word-vec and random forest	Daily	Accuracy	70.1%
Gao et al. [19]	SemEval-2014 task 4 (restaurant and laptop) and tweets	BERT-pair-QA-MUL	Long time	Accuracy and micro F1	89.27%, 79.15%

vocabulary lexicon is generated. In addition, this paper constructs a Long Short-Term Memory (LSTM) model [24] with Python open source package. Based on the lexicon built above, this study choses 23,152 comments as the test data, and evaluates the sentiment analysis models. The correlation coefficient and Apriori algorithm are utilized to discuss the relationship between investor sentiment and the stock market trend. The code programs involved in this research has been uploaded to Github for future research.

3.1 Dataset

Stock forums such as Sina and Easymoney.com stock forum are online platforms for Chinese stockholders to express their opinions [25]. They are the main data sources for studying the investor focuses and emotions in China's stock market. This study uses Python programming language to crawl the investor generated contents on the two platforms. Stock data involves the "date", "opening", "closing", "highest", "lowest", "volume", "amplitude", and "exchange" information has been collected, the crawler coded in python is uploaded to Github: https://github.com/luckanny111/stock_crawler_once.git, and https://github.com/luckanny111/sina_guba_crawler.git. A technique of WordCloud is employed to present the major information of crawled dataset. Python codes of WordCloud figure imaging is shared at https://github.com/luckanny111/ciyun.git.

3.2 SentiCon Building

This study extracts the 3000 keywords based on the Term Frequency-Inverse Document Frequency (TF-IDF) algorithm and TextRank algorithm to. The first 2000 words are selected as candidate words to build the stock sentiment lexicon. After that, the article matches the candidate words with Tsinghua University Chinese Complimentary Dictionary, National Taiwan University NTUSD Sentiment Dictionary, and HowNet Sentiment Dictionary, and uses the method of manually labeling sentiment scores to build the SentiCon based on the candidate words. The process can be divided into three steps. First, automatic extract the candidate words. Second, match candidate words with the universal language polarity database. Third, manual label sentiment scores to improve the dictionary. TextRank keyword extraction is able to be implemented directly by calling the sklearn module function, and the TF-IDF algorithm Python code has been uploaded to Github: https://github.com/luckanny111/IF-IDF.git. The calculation formula of the TF-IDF value of the keyword in the document is as follows:

$$TF_{(w,D_i)} - IDF_{(w,D_i)} = \frac{count(w)}{|D_i|} \times \log \frac{N}{(1 + \sum_{i=1}^{N} I(w, D_i))}, \quad (1)$$

The iterative calculation formula of TextRank is as follows:

$$WS(V_i) = (1 - d) + d \times \sum j \in In(V_i) \frac{w_{jk}}{\sum (V_i \in Out(V_j))_{jk}^{w}} WS(V_i), \quad (2)$$

Comparing TF-IDF and TextRank algorithms, both of them are performed on the result of word segmentation. Therefore, whether to add labeled keywords into the corpus causes a huge difference in accuracy and recall rates. Therefore, considering of different advantages of the two algorithms, this paper utillizes the TF-IDF and TextRank algorithm to extract 3000 word-level keywords individually, and selects the first 2000 words of the intersection of two results as the Stock SentiCon candidate words. With 2000 candidate words, this study labels words sentimental polarities by matching them with common SentiCons, which includes Tsinghua University Chinese sentiment lexicon (http://nlp.csai.tsinghua.edu.cn/lj/sentiment.dict.v1.0.zip), National Taiwan University Sentiment Dictionary (NTUSD) (http://nlg18.csie.ntu.edu.tw:8080/opinion/index.html) and HowNet sentiment lexicon (http://keenage.com/). These lexicons are available to be obtained from the internet.

They are also be utilized and verified in emotional orientation recognition tasks [26, 27]. Furthermore, sentiment polarity are artificially proofread, a polarity and a weight was involved in the lexicon. The polarity is recoded as "−1" (negative), "0" (neutral) or "1" (positive), weights are decided by TF-IDF and TextRank algorithms.

3.3 *Sentiment Analysis*

3.3.1 Lexicon-Based Sentiment Analysis

The construction of the sentiment analysis model based on the sentiment dictionary mainly includes the constructing the Sentiment lexicon, polarity labeling words, calculating sentiment scores and testing the model. Lexicon based sentiment analysis method is mainly divided into five steps.

1. Perform word segmentation processing of the document.
2. Traverse the word segmentation results, locate the emotional words according to the sentiment lexicon objects imported in the second step, and judge whether there are degree adverbs and negative words between the emotional words.
3. The initial weight of sentiment word is assigned a value of 1, if a negative word appears, the next emotional word weight value is reversed.
4. Calculate emotional word scores.
5. Calculate the document score, which is the sum of the scores of all sentiment words in the article.
6. Judge the sentiment polarity as positive (recorded as 1), and negative (recorded as 0).

The code of lexicon-base sentiment analysis model shared on Github: https://github.com/luckanny111/sentiment_analysis_lexicon.git.

3.3.2 Sentiment Analysis Based on LSTM Algorithm

LSTM is a special Recurrent Neural Network (RNN), each node is connected to form a loop and recursively to achieve classification and prediction tasks [23]. RNN can simulate human reading sequence, read serialized data and transmit information through coding and memory of hidden layer neurons. However, because ordinary RNNs can only remember short-term sequence data and cannot solve the problem of long-term dependence, the phenomenon of gradient disappearance will appear. Figure 1 indicates that LSTM is an extended version of RNN, adding three control units (namely input gate, output gate and s forget gate) and a storage unit. When the information enters the model, the elliptic curve will judge the information, the information that meets the rules will be retained, otherwise, the information will be forgotten, and the timely memory function is realized through the door to prevent the slope from disappearing. This paper applies the Python programming language

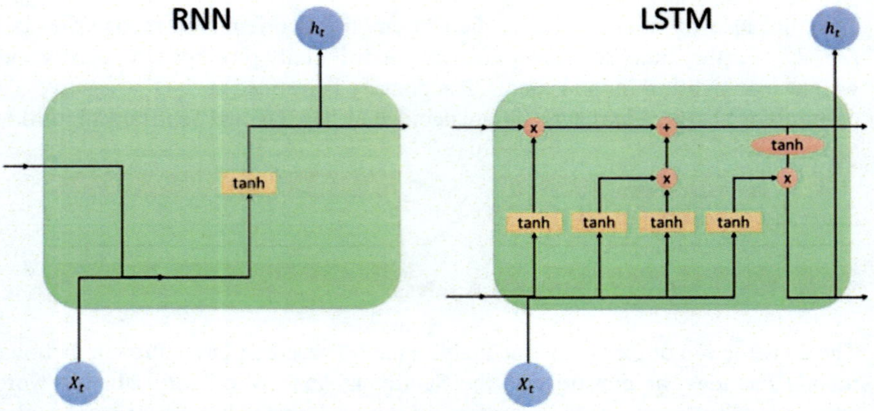

Fig. 1 The architecture framework of LSTM and RNN

to implement the construction of the LSTM model. Select 71,775 pieces of data as the training data set and the verification data set. The size of negative data set is 55769 and the positive data set is 16006. The training model uses Sigmoid as the activation function, Adam as the optimization function, and the number of iterations is set as 10. The Python code of LSTM model is shared at github: https://github.com/luckanny111/LSTM_sentiment_analysisi.git.

3.4 Correlation Analysis

This paper uses correlation coefficient and Apriori algorithm to realize the correlation research between stock trend and stockholder sentiment. The stock trend indicator is determined by the stock trend of the day, namely: Stock Trend = Closing Price-Opening Price, if Stock Trend > 0, the trend of the day is up, recorded as "1", otherwise recorded as "0". Apriori is a classic correlation algorithm, where the most famous case is "Beer and Diapers" in the United States. The correlation coefficient calculation is implemented by calling the numpy module of python, and the code of Apriori algorithm is shared by uploading to Github: https://github.com/luckanny111/correlation_aproori_corr.git. The calculation formulas for the two indicators of Apriori (Support, Confidence) and correlation coefficient are as follows:

$$Support(X => Y) = P(X \cup Y)\ , \tag{3}$$

$$Confidence(X => Y) = P(X/Y)\ , \tag{4}$$

$$r(X, Y) = \frac{Cov(X, Y)}{\sqrt{Var[X]VAR[Y]}}\ , \tag{5}$$

4 Results and Discussions

This article applies crawler technology to crawl a total of 530,023 stockholders' comments. With 23,152 user comments are selected for model measuring. Results show that more than 50% online investor comments express negatively. Comparing accuracies of the two models, the sentiment analysis model of the LSTM algorithm shows a bigger advantage, where the accuracy of the LSTM algorithm is 99.87%, and the accuracy of the model based on the sentiment dictionary is 94.57%. In addition, as Table 4 shown, through analyzing the delay effect on the influence of sentimental tendencies and stock trends, the current sentiment score has the highest impact on the trend of stocks on the third day afterwards. The Apriori support rate is 0.310758, the confidence rate is 0.557064, and the correlation coefficient is 0.008642. All of the research results are available at Github: https://github.com/luckanny111/sentiment_stock_results.git.

4.1 Data Collection Results

530,023 pieces of comment are obtained from stock forum, where 44,087 and 485,936 pieces of information are generated in Sina and Fortune.com forums respectively. Pertaining to crawled results, this study builds a 530,023-data-level economic corpus and presents a descriptive analysis. Table 3 shows sentiment results based on lexicon-based and LSTM-based sentiment analyzing approaches.

This paper presents a word cloud plotting figure based on the TF-IDF algorithm as Fig. 2. This figure indicates that keywords of "Financing", "securities lending", "information", "Huaxia", "Guizhou", "Maotai", "capital", "trading" and "RMB common stock" maybe the major topics of investors, words of "daily trading limit" and "garbage" maybe sentiment words, which describe positive and negative feelings respectively.

Table 3 Sentiment analysis results

Data resource	Number of pieces	Lexicon-based method		LSTM-based method	
		Negative	Positive	Negative	Positive
Fortune stock forum	485,936	346506	139430	293150	192707
Sina stock forum	44,087	17742	26345	16703	27384
The sum	530,023	364248	165775	09853	220091

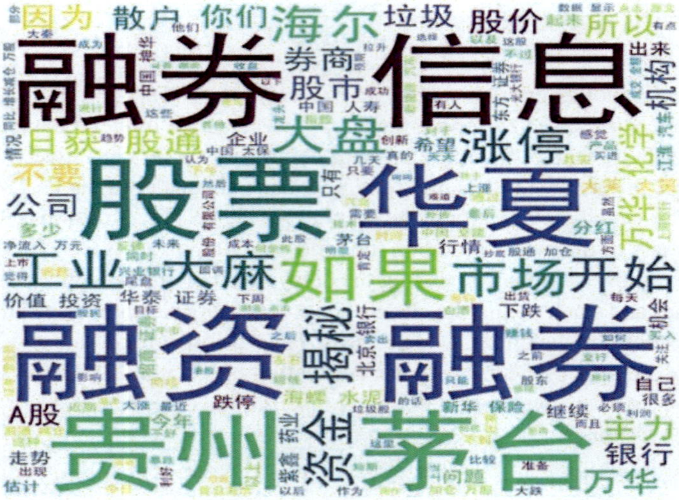

Fig. 2 Topic visualization with word-cloud figure

4.2 Results of Sentiment Lexicon Building

This study applies TF-IDF algorithm and TextRank algorithm to extract keywords as candidate words. Select the intersection of results of the two algorithms, this research extract 2000 sentiment candidate words, top 50 of which are: “financing”, “margin trading”, “stock”, “Moutai”, “hemp”, “company”, “market”, “Guizhou”, “automobile”, “share”, “share price”, “Baijiu”. “Jianghuai”, “capital”, “plate”, “market”, “stock”, “investment”, “Zixin”, “industry”, “science and technology”, “industry”, “securities”, “China”, “enterprise”, “transaction”, “net inflow”, “market”, “pharmaceutical”, “rise”, “performance”, “product”, “present”, “technology”, “individual stock”, “New energy”, “trading limit”, “ginseng”, “index”, “price”, “trading limit”, “Hisense”, “quotation”, “market value”, “stock market”, “development”, “increase”, “everyone”, “buy” and “growth”. This result shows a high similarity with word cloud result. After comparing with general sentiment lexicons and artificially labeling processing, a sentiment lexicon is built with 2000 words involved, for instance, “risk” and “down” represents a negative sentiment polarity, “deal”, “trade”, “up” presents a positive emotion. Partial results are presented as Table 4.

4.3 Sentiment Analysis Results

Sentiment lexicon and LSTM based methods are applied to perform sentiment analysis with 530,023 pieces of comments. Regarded 23,152 pieces of data as the test data

Table 4 Partial results of sentiment lexicon and corpus

Words	Sentiment	Words	Sentiment	Words	Sentiment
Deal	1	Daily	Trading limit	1	R&D
0					
Net inflows	0	Rise	1	Down	−1
Market	0	People	0	Stock	0
Pharmaceutical industry	0	Buy	0	Trend	
0					
Up	1	Increase	1	Risk	−1

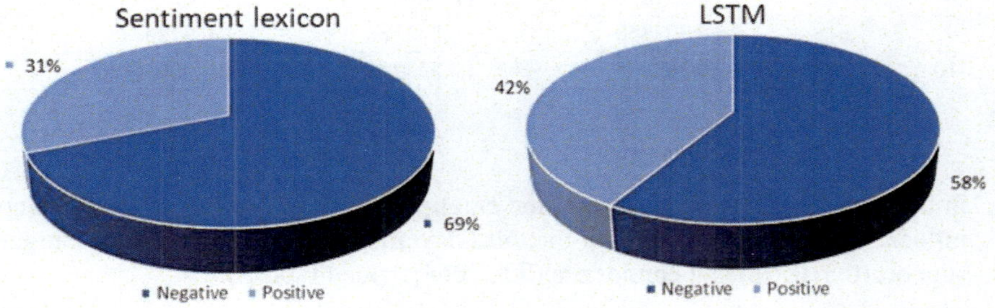

Fig. 3 Sentiment analysis results

set, LSTM based method presents 99.87% accuracy, which is higher than lexicon-based approach. Results of the lexicon method indicates that 31% of the comments are positive and 69% are negative. Results based on LSTM model show that 42% of reviews are positive and 58% are negative. Thus, this concludes that there are more negative emotions in online reviews (Fig. 3).

4.4 Correlation Analysis

This paper performs a correlation analysis based on LSTM sentiment analysis results of 23,152 pieces of data. Correlations between the investors emotion and stock trend are explored by applying the correlation coefficient parameter and Apriori algorithm. Considering the delay of the influence of investor emotions on the stock trend, this paper performs correlation analysis on current sentiment results with stock trends

Table 5 Analysis results of correlation between investors' sentiments and stock trend

Days	Apriori		Correlation coefficient
	Support	Confidence	
0	0.298024	0.548501	0.057033
1	0.302139	0.547131	0.015759
2	0.302087	0.545992	0.006369
3	0.310758	0.557064	0.008642
4	0.290577	0.524524	0.023423
5	0.298072	0.533816	0.011682
6	0.301246	0.546003	0.002202
7	0.30447	0.555038	0.023111
8	0.299611	0.546701	0.011594
9	0.303358	0.552522	0.012677
10	0.309205	0.55394	0.009998

in the next 10 d. Results show that the emotional tendency of the day has a deeper influence on the stock market of the third day afterwards, which presents a highest support (0.310758) and confidence (0.557064) value (Table 5).

5 Conclusion

Regarding to China's stock market, more than 80% are individual stock owners, they are also the major generators of finance social media contents. By studying the effect of online investor emotions on stock trends, this study delivers helpful insights for investors with decision-making supports. Taking 20 stocks of the Shanghai Stock Exchange from 2017 to 2019 as an example, this study presents the methods of web crawling, keyword extraction, word cloud plotting, finance SentiCon building, sentiment analyzing based on lexicon and LSTM, and correlation analysis. Lexicon-based and LSTM-based sentiment analysis methods are compared, where LSTM shows a great advantage with it high accuracy. The relationship between investor emotions and stock markets is discussed by applying Apriori and correlation coefficient approaches. Findings show that current stockholder emotion is more correlated with the stock fluctuations on the third day afterwards. Furthermore, codes in Python program language of mentioned algorithms are contributed in authors' Github. Limitations of this paper is that, as far as the correlation analysis results are concerned, the delay effect difference between the dates is still weak to be noticed. Thus, the article raises following suggestions for future researches.

1. Consider multiple unstructured data as the data source for sentiment analysis, such as pictures, videos, symbols, and facial expressions.
2. Conduct experiments on various sentiment analysis models with diversified polarities.
3. Apply different algorithms to perform correlation analysis and explore the relationships between stockholders' emotions and stock trends.
4. Utilize varied stock and sentiment indicators with massive data sets to explore insights of relationships between investor emotions and finance markets.

Acknowledgements This work is in part supported by the national key research project [YFE0101000], 2020 Key Technology R&D Program of GuangDong Province ZH01110405180056PWC] and Zhuhai Technology and Research Foundation [TC200802D4]. Thanks for the funding of mentioned projects.

References

1. Li, D., Wang, Y., Madden, A., Ding, Y., Tang, J., Sun, G.G., Zhang, N., Zhou, E.: Analyzing stock market trends using social media user moods and social influence. J. Assoc. Inf. Sci. Technol. **70**(9), 1000–1013 (2019)
2. Aggarwal, U., Saxena, A., Herald, S.: Artificial intelligence review in stock markets. Int. J. Res. Eng. Sci. Manag. **2**(11), 92–95 (2019)
3. Vachhani, H., Obiadat, M.S., Thakkar, A., Shah, V., Sojitra, R., Bhatia, J., Tanwar, S. : Machine learning based stock market analysis: a short survey. In: International Conference on Innovative Data Communication Technologies and Application, pp. 12–26. Springer, Cham (2019)
4. Rossi, M., Gunardi, A.: Efficient market hypothesis and stock market anomalies: empirical evidence in four European countries. J. Appl. Bus. Res. (JABR) **34**(1), 183–192 (2018)
5. Kumar, H., Jawa, R.: Efficient market hypothesis and calendar effects: empirical evidences from the Indian stock markets. Bus. Analyst **37**(2), 145–160 (2017)
6. Shah, D., Isah, H., Zulkernine, F.: Stock market analysis: a review and taxonomy of prediction techniques. Int. J. Finan. Stud. **7**(2), 26 (2019)
7. Dash, M.: Testing the random walk hypothesis in the Indian stock market using ARIMA modelling. J. Appl. Manag. Investments **8**(2), 71–77 (2019)
8. Nasr, N., Farhadi Sartangi, M., Madahi, Z.: A fuzzy random walk technique to forecasting volatility of Iran stock exchange index. Adv. Math. Finan. Appl. **4**(1), 15–30 (2019)
9. Shaik, M., Maheswaran, S.: Random walk in emerging Asian stock markets. Int. J. Econ. Finan. **9**(1), 20–31 (2017)
10. Liu, B., Zhang, L.: A survey of opinion mining and sentiment analysis. In: Mining Text Data, pp. 415–463. Springer, Berlin (2012)
11. Pang, B., Lee, L.: Opinion Mining and Sentiment Analysis (Foundations and Trends (R) in Information Retrieval). Now Publishers Inc. (2008)
12. Schumaker, R.P., Chen, H.: A quantitative stock prediction system based on financial news. Inf. Process. Manag. **45**(5), 571–583 (2009)
13. Schumaker, R.P., Chen, H.: Textual analysis of stock market prediction using breaking financial news: the AZFin text system. ACM Trans. Inf. Syst. (TOIS) **27**(2), 12 (2009)
14. Bollen, J., Mao, H., Zeng, X.: Twitter mood predicts the stock market. J. Comput. Sci. **2**(1), 1–8 (2011)
15. Mittal, A., Goel, A.: Stock Prediction Using Twitter Sentiment Analysis. Standford University, CS229 (2012). Available online. http://cs229.stanford.edu/proj2011/GoelMittalStockMarketPredictionUsingTwitterSentimentAnalysis.pdf. Cited 23 June 2021

16. Lee, H., Surdeanu, M., MacCartney, B., Jurafsky, D.: On the importance of text analysis for stock price prediction. In: The 9th International Conference on Language Resources and Evaluation. LREC 2014, pp. 26–31. Reykjavik, Iceland (2014)
17. Kalyanaraman, V., Kazi, S., Tondulkar, R., Oswal, S.: Sentiment analysis on news articles for stocks. In: The 2014 8th Asia Modelling Symposium (AMS), pp. 23–25. Taipei, Taiwan (2014)
18. Cakra, Y.E., Trisedya, B.D.: Stock price prediction using linear regression based on sentiment analysis. In: The 2015 International Conference on Advanced Computer Science and Information Systems (ICACSIS), pp. 10–11. Depok, Indonesia (2015)
19. Gao, Z., Feng, A., Song, X., Wu, X.: Target-dependent sentiment classification with BERT. IEEE Access **7**(1), 154290–154299 (2019)
20. Pagolu, V.S., Reddy, K.N., Panda, G., Majhi, B.: Sentiment analysis of Twitter data for predicting stock market movements. In: The 2016 International Conference on Signal Processing, Communication, Power and Embedded System (SCOPES), pp. 3–5. Paralakhemundi, India (2016)
21. Xu, Y., Cohen, S.B.: Stock movement prediction from tweets and historical prices. In: The 56th Annual Meeting of the Association for Computational Linguistics, pp. 15–20. Melbourne, Australia (2018)
22. Mohammed, M., Omar, N.: Question classification based on Bloom's taxonomy cognitive domain using modified TF-IDF and word2vec. PloS One **15**(3) (2020). https://doi.org/10.1371/journal.pone.0230442
23. Kazemi, A., Pérez-Rosas, V., Mihalcea, R.: Biased: TextRank: Unsupervised Graph-Based Content Extraction (2020). arXiv preprint arXiv:2011.01026. Available online. https://arxiv.org/pdf/2011.01026.pdf. Cited 23 June 2021
24. Ombabi, A.H., Ouarda, W., Alimi, A.M.: Deep learning CNN-LSTM framework for Arabic sentiment analysis using textual information shared in social networks. Soc. Network Anal. Min. **10**(1), 1–13 (2020)
25. Beckman, M.D., Çetinkaya-Rundel, M., Horton, N.J., Rundel, C.W., Sullivan, A.J., Tackett, M.: Implementing version control with Git and GitHub as a learning objective in statistics and data science courses. J. Stat. Educ. **29**(Sup 1), 1–35 (2020)
26. Bo, Y., Liu, Y., Li, H.: Sentiment classification in Chinese microblogs: lexicon-based and learning-based approaches. Int. Proc. Econ. Dev. Res. **68**(1), 1–5 (2013)
27. Fulian, Y., Wang, Y., Liu, J., Lin, L.: The construction of sentiment lexicon based on context-dependent part-of-speech chunks for semantic disambiguation. IEEE Access **8**(1), 63359–63367 (2020)

A Framework and Decision Algorithm to Determine the Best Feature Extraction Technique for Supporting Machine Learning-Based Hate Speech Detection

Chun-Kit Ngan and Kashyap Bhuva

Abstract We develop and implement a framework and a decision algorithm to determine the best feature extraction technique (FET) for supporting machine learning-based hate speech detection. Specifically, the contributions of this work are three-fold: (1) a seamless modular pipeline that automatically preprocesses, vectorizes, and classifies whether or not a text message is a hate speech; (2) a decision algorithm that determines the best FET approach among all the possible FET candidates with the linear time complexity *O(N)*; and (3) a preliminary experimental evaluation on the tweets provided by Twitter Sentiment Analysis on Analytics Vidhya to demonstrate that our FET framework and decision algorithm are effective and produce the significant results.

Keywords Feature extraction framework · Decision algorithm · Hate speech detection · Natural language processing · Machine learning classifier

1 Introduction

In the Big Data era, there are large numbers of text-based messages available on social media and online community forums such as Facebook, YouTube, and Twitter. These virtual communication forums enable anonymous users to express themselves freely and irresponsibly that disparage, offend, insult, intimidate, or threaten an individual or a group based upon some of their characteristics including race, color, ethnicity, gender, sexual orientation, nationality, region, religion, disability, age, or other characteristics [1–3]. Some examples that we found from [4, 5] are "Wipe out the Jews", "Women are like grass, they need to be beaten/cut regularly", "The Palestinians are beasts walking on two legs", "Don't trust boys", "I hate migrants",

C.-K. Ngan (✉) · K. Bhuva
Data Science Program, Worcester Polytechnic Institute, 100 Institute Road, Worcester, MA 01609, USA
e-mail: cngan@wpi.edu

K. Bhuva
e-mail: kbbhuva@wpi.edu

R. Lee (ed.), *Computer and Information Science 2021—Summer*, Studies in Computational Intelligence 985,
https://doi.org/10.1007/978-3-030-79474-3_2

and more. These kinds of message contents are called hate speech that targets and causes harm to identifiable groups and individuals, with or without malicious intent [6]. Note that these examples found on the Internet and documents do not represent the authors' viewpoints. Thus, hate speech detection has become an important process for analyzing people's sentiment, i.e., a user or a group against another user or group, and for discouraging associated wrongful activities.

To support hate speech detection, supervised machine learning-based (ML) classification models have been widely used. Three of the most popular algorithms being used in this area include Support Vector Machine (SVM) [7, 8], Naïve Bayes (NB) [9, 10], and Random Forest (RF) [11, 12]. These algorithms can construct the models, based upon the feature characteristics from a large number of text-based messages, to detect hate speech. Presently, the most effective methods that can extract and select the most important features from hate speech can be broadly divided into two distinct word-embedded categories: (1) Statistical-Based and (2) Neural-Network-Based.

The former method is a keyword- and keyphrase-based approach, namely Term Frequency–Inverse Document Frequency (TF-IDF) [13, 14]. This approach considers the frequency of a word or a set of words (i.e., n-grams, where $n \geq 1$), as a token, within a single document and its inverse frequency in an entire text corpus. It means that a token with a larger number of occurrences within the same document and a fewer number of occurrences across the entire text corpus has a higher feature value compared with that of the other tokens. The advantages of this approach are two-fold: (1) the process to compute this basic feature metric of each token is simple and easy that we can extract the most descriptive terms in a document and (2) the procedure to measure the similarity between any two documents is fast and effective that we can use the cosine similarity score [15] to identify the relevant document pair. However, the TF-IDF is based upon the Bag-of-Words (BoW) or the Bag-of-N-Grams (BoNG) model that does not capture word positions, semantics, and co-occurrences across different documents and is useful only in lexical level features.

The latter approach is a semantic-based representation which maps related words to a vector based upon the context of the text corpus. One of the most popular techniques that is often used is Word2Vec (W2V) [16, 17]. There are two different approaches to train W2V embeddings: (1) Continuous Bag of Words (CBoW) that learns and predicts the target word from the adjacent words and (2) Skip-Grams (SG) that is designed to learn and predict the adjacent target words from a word of interest. According to [18], CBoW is several times faster to be trained and has a slightly better accuracy for frequent words. SG work even well with a small amount of training data and represent well rare words or phrases. In this pilot study, we mainly focus on CBoW for the experimental observation and evaluation.

Apparently, none of the aforementioned approaches, N-grams TF-IDF and CBoW W2V, can be claimed to be the leading role to support the hate speech detection. To answer this research question, we develop and implement a framework and a decision algorithm to determine the best feature extraction technique (FET) for supporting ML-based hate speech detection. This framework is a seamless modular pipeline that automatically preprocesses, vectorizes, and classifies if a text message is a hate speech, from which to determine the best FET approach among all the possible candidates for each real-case scenario. Specifically, the pipeline first takes the text messages from the corpus and then preprocesses the text, including sentence splitting, spelling correction, contraction expanding, punctuation removals, etc., to generate the cleaned text. After that, the cleaned text is sent to the vectorizer to generate a vector by using both N-grams TF-IDF and CBoW W2V approaches. As the inputs, the N-grams TF-IDF and CBoW W2V vectors are then passed to the classifier module to construct the ML-based models, including NB, SVM, and RF, on the training dataset. The trained models are then evaluated on the validation dataset to determine which feature extraction approach, N-grams TF-IDF or CBoW W2V, delivers a better detection performance in terms of accuracy, precision, recall and F1 score by using our developed decision algorithm. To demonstrate the effectiveness of our framework and decision algorithm, we conduct a preliminary experimental study (i.e., 31,962 tweets provided by Analytics Vidhya [19]) and evaluate the results on the testing dataset to conclude that the CBoW W2V model does not always outperform the N-grams TF-IDF approach on hate speech detection. Note that the tweets provided by Analytics Vidhya are stored in a csv file, in which each row has three attributes, i.e., ID, Label (1—Hate and 0—Like), and Content, for each tweet.

The remainder of the paper is organized as follows. First, we describe our framework for tweet preprocessing, vectorization, and classification in Sect. 2. Section 3 explains our decision algorithm for selecting the best FET approach among all the possible candidates. In Sect. 4, we conduct the experimental study, illustrate the results, and conclude the discussion in our study. In Sect. 5, we summarize our contribution and briefly outline our future work.

2 FET Framework

Figure 1 on the next page is our FET framework that is a seamless modular pipeline that consists of three main modules: Preprocessor, Vectorizer, and Classifier. Each module has sub-components that manipulate and process texts.

2.1 Tweet Preprocessor

Tweet Preprocessor is composed of 11 sub-modules that includes Tweet Separator (TS), Sentence Splitter (SS), Spelling Corrector (SC), Contraction Expander (CE), Punctuation Remover (PR), Non-alphanumeric Remover (NR), Stopword Remover (SR), Emoji Remover (ER), Hashtag Remover (HR), Word Lemmatizer (WL), and Lowercase Converter (LC).

First, Text Preprocessor takes a tweet corpus, i.e., a collection of 31,962 tweets, as an input and the TS module segregates them from the corpus into many individual tweets. Each tweet is then sent to the SS module, which splits the tweet into individual sentences. The SC module conducts the spell check on each sentence

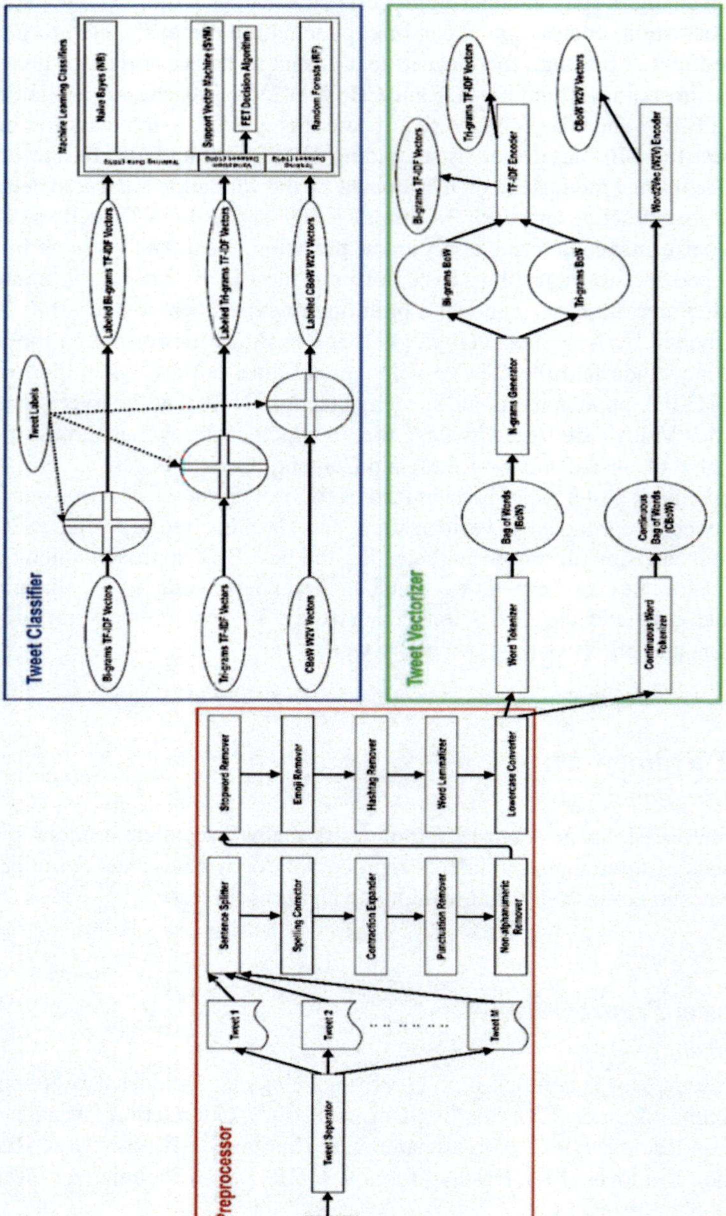

Fig. 1 FET framework

and uses the unigram tokenization approach to replace misspelled words with the highest-probability corrected words. The corrected-word sentences are then sent to the CE module, which expands the contracted form of the words into a longer form, such as "I'm" to "I am", "You're" to "You are", and "It's" to "It is", in each sentence. The subsequent modules, PR, NR, SR, ER, and HR, respectively removes punctuations (e.g., full stop (.), comma (,), and colon (:)), non-alphanumeric characters (e.g., #GoLangCode123!$! to GoLangCode123), stopwords (e.g., "ourselves", "hers", "between", "yourself", "but", and "again"), emojis (e.g., 😂, 😫, 😢, and 😱), and hashtags (e.g., #coffeelovers, #COVID-19, and #TheOscars) from expanded sentences. To reduce the dimension of each expanded sentence, the WL module groups several forms of the same meaning word together even if their spellings are quite different. For example, "walk", "walked", "walks", and "walking" would be treated the same as "walk". This extensive normalization down to the semantic root of a word is called lemmatization. Finally, the LC module converts each character of a word into a lowercase on lemmatized sentences and then sends the processed sentences to Tweet Vectorizer.

2.2 *Tweet Vectorizer*

Tweet Vectorizer is formed by two encoding pipelines: N-grams TF-IDF and CBoW W2V. Both pipelines take the preprocessed tweets as the inputs to generate the corresponding tweet vectors.

The N-grams TF-IDF pipeline includes Word Tokenizer (WT), N-grams Generator (NG), and TF-IDF Vectorizer. First, the preprocessed tweets generated from Tweet Preprocessor are passed into the WT module that splits each sentence of a tweet into a collection of single vocabularies called tokens or unigrams, i.e., Bag of Words (BoW). It is called BoW, as the positional order and the structure of those vocabularies are discarded but only their occurrences in the document are annotated. The NG module then creates a keyphrase that is based upon adjacent words that are grouped together. For the example of bi-gram BoW, "This is a sentence" has the following 2-grams: (this is), (is a), and (a sentence). For the instance of tri-gram BoW, the same sentence has the following 3-grams: (this is a) and (is a sentence). Both n-grams BoW are then sent to the TF-IDF Encoder to generate bi-gram and tri-gram TF-IDF vectors.

The CBoW W2V is composed of two sub-modules: Continuous Word Tokenizer (CWT) and W2V Encoder (WE). The CWT module tries to predict the current center word, as a target, based upon the surrounding words from a sentence. For example, suppose that we consider a sentence "I like Natural Language Processing" with the target word (i.e., "Natural") and the size of the sliding window equal to 2. The CWT maps the word "Natural", as the target output, based on the two adjacent words on the left (i.e., "I" and "like") and the two adjacent words on the right (i.e., Language

Processing), as the inputs, to generate a CBoW. Using the input–output pairs, the WE module that is a deep neural network to generate the CBoW W2V vectors.

2.3 Tweet Classifier

Tweet Classifier consists of two sub-modules: Concatenators and ML Classifiers. The role of each concatenator is to combine the vectors generated from Tweet Vectorizer and the corresponding target labels (i.e., Hate (1) or Like (0)) to generate the labelled vectors with the unique identification number, which are then passed to the ML classifiers. For each set of labelled vectors, they are randomly separated into three portions: training dataset (80%) to construct three models (i.e., NB, SVM, and RF); validation dataset (10%) to evaluate each model performance and select the best FET approach, using our developed decision algorithm, in terms of accuracy and F1 score; and testing dataset (10%) to demonstrate the effectiveness of our pipeline and algorithm.

3 FET Decision Algorithm

In this section, we discuss in detail the decision algorithm used in selecting the best FET approach among all the possible candidates. More specifically, we need to define a set of mathematical notations for it shown in Table 1 on the next page.

The decision algorithm to solve the BFET problem is generalized in the following three steps: (1) check with BDTT if ODM is a balanced dataset; (2) compute the average accuracy and F1 score on VDPM among all the CMs for each FET; and (3) return the BFET based upon the average accuracy and F1 score among all the possible FETs. The pseudo code of the algorithm with the detailed comments is further explained and shown in Table 2 on Page 8.

4 Experimental Results and Discussion

The total time complexity of this decision algorithm is $O(N + n * m)$, where is N is the total number of data instances of ODM, n is the total number of data instances of VDPM, and m is the total number of unique FETs. In our experimental study, N is 31,962, n is 9, and m is 3. Since N is much larger than $n * m$ (i.e., 27), the time complexity of this algorithm is $O(N)$. Using the algorithm for this given imbalanced dataset, the BFET is Bi-grams TF-IDF on VDM in terms of Accuracy and F1 score. To demonstrate the effectiveness of our framework and decision algorithm, we perform the detection on the testing dataset to evaluate the results shown in Table 3 on Page 13.

Table 1 FET mathematical notations

Definition	Description
1	**Original Data Matrix (ODM)**: An ODM is a two-dimensional array matrix that stores the original message vector dataset with three positional attributes: Unique Id, Class Label, and Message Vector
2	**Validation Data Matrix (VDM)**: A VDM is a two-dimensional array matrix that stores 10% of the ODM dataset for evaluating the classifiers' performance
3	**Validation Data Performance Matrix (VDPM)**: A VDPM is a two-dimensional array matrix that stores the classifiers' performance on VDM. Specifically, the VDPM includes five positional attributes: Feature Extraction Technique (FET), Classification Model (CM), Accuracy (Acc), Precision (Pre), Recall (Re), and F1 score (F1)
4	**Balance Dataset Tolerance Threshold (BDTT)**: A BDTT is a tolerance percentage value between 0 and 10% determined by users to decide if ODM is a balanced dataset
5	**Best Feature Extraction Technique (BFET)**: A BFET is a variable that stores the best feature extraction technique among all the FET candidates
6	**BFET Problem and Solution**: A BFET problem is a tuple *<ODM, VDPM, BDTT>*, in which a solution to the problem is a BFET that can maximize the average Accuracy and F1 score on VDM among all the possible FET candidates

From the result, we can see that Bi-grams TF-IDF is still the best in terms of Accuracy (92.33%), Precision (76.67%), Recall (78.33%), and F1 score (75.33%).

5 Conclusion and Future Work

To our best knowledge, it is the first paper to develop and implement a framework to determine the best feature extraction technique for supporting machine learning-based hate speech detection. Specifically, the contributions of this work are three-fold: (1) a seamless modular pipeline that automatically preprocesses, vectorizes, and classifies whether or not a text message is a hate speech; (2) a decision algorithm that selects the best FET approach among all the possible FET candidates with the linear time complexity *O(N)*; and (3) a preliminary experimental evaluation on the tweets provided by Twitter Sentiment Analysis on Analytics Vidhya to demonstrate that our decision framework and algorithm are effective and produce the significant results. However, there is still a lack of many important research questions, e.g., how other FET approaches could be integrated into the decision framework, what other performance metrics should be used for the FET selection, how the proposed algorithm could be enhanced to reduce the time and space complexity, and what other available datasets should be chosen for the performance evaluations.

Table 2 FET decision algorithm

Input: *<ODM, VDPM, BDTT>*
Output: *<BFET>*

Initialization:

```
// Boolean Variables
balanceDataset = False; // Indicate if the dataset is balanced or not.

// Integer Variables
numOfZero = 0; // Store the total number of "Like" messages
numOfOne = 0; // Store the total number of "Hate" messages
sizeOfDataSet = 0; // Store the total number of rows of ODM
maxF1ScoreIndex = 0; // Store the maxF1Score index
maxAccIndex = 0; // Store the maxAccScore index
sizeOfUniqueFETArray = 0; // Store the total number of FETs
sizeOfVDPM = 0; // Store the total number of rows of VDPM

// Float Variables
numOfZeroPercent = 0; // Store the total percentages of "Like" messages
numOfOnePercent = 0; // Store the total percentages of "Hate" messages
maxF1Score = 0; // Store the max average F1 score among all the FETs'
maxAccScore = 0; // Store the max average Acc score among all the FETs'
```

(continued)

Table 2 (continued)

// Array Variables
countOfCM = []; // Store the total number of CMs for each FET
sumOfAcc = []; // Store the total accuracy of CMs for each FET
sumOfF1 = []; // Store the total F1 score of CMs for each FET
avgOfAcc = []; // Store the average accuracy of CMs for each FET
avgOfF1 = []; // Store the average F1 score of CMs for each FET
uniqueFETArray = []; // Store each unique FET

Processing:
 	
<u>STEP 1</u>: Check with BDTT if ODM is a balanced dataset.
 	
// Store the total number of rows of ODM
sizeOfDataSet = len(ODM);
 	
// Compute the total number of "Like" & "Hate" messages in ODM
FOR (i := 1; i <= sizeOfDataSet; i := i + 1) **DO** {
  IF (ODM[i][2] == 0) **THEN**
    numOfZero = numOfZero + 1;
  ELSE
    numOfOne = numOfOne + 1;
  ENDIF}

(continued)

Table 2 (continued)

```
// Compute the total percents of "Like" & "Hate" messages in ODM
numOfZeroPercent = numOfZero/sizeOfDataSet;
numOfOnePercent = numOfOne/sizeOfDataSet;

// Check if ODM is a balanced dataset
IF |numOfOnePercent - numOfZeroPercent| <= BDTT THEN
   balanceDataset = True
ENDIF

STEP 2: Compute the average Accuracy and F1 score on VDPM among all the classification models for each FET.

// Store the unique FET techniques to the array
uniqueFETArray = VDPM.FET.unique();
// Store the total number of FET techniques
sizeOfUniqueFETArray = len(uniqueFETArray);
// Store the total number of rows of VDPM
sizeOfVDPM = len(VDPM);

// Assign the initial zero value to each element in the array
FOR (i := 1; i <= sizeOfUniqueFETArray); i := i + 1) DO {
   countOfCM.append(0);
   sumOfAcc.append(0);
   sumOfF1.append(0);
   avgOfAcc.append(0);
   avgOfF1.append(0);}
```

(continued)

Table 2 (continued)

```
// Compute the total number of classification models, the total sum of accuracy, and the total sum of F1 score for each FET ap-
proach
FOR (i := 1; i <= sizeOfVDPM; i := i + 1) DO {
  FOR (j := 1; j <= sizeofUniqueFETArray; j := j + 1) DO {
    IF (VDPM[i][1] == uniqueFETArray[j]) THEN
      countOfCM[j] = countOfCM[j] + 1;
      sumOfAcc[j] = sumOfAcc[j] + VDPM[i][3];
      sumOfF1[j] = sumOfF1[j] + VDPM[i][6];
    ENDIF}}

// Compute the average of accuracy and F1 score among all the classification model for each FET approach
FOR (i := 1; i <= sizeOfUniqueFETArray; i := i + 1) DO {
  avgOfAcc[i] = sumOfAcc[i] / countOfCM[i];
  avgOfF1[i] = sumOfF1[i] / countOfCM[i];}

STEP 3: Return the best feature extraction technique based upon the average Accuracy and F1 score among all the FETs.

// Find the max average accuracy and F1 score among all the FETs
FOR (i := 1; i <= sizeOfUniqueFETArray; i := i + 1) DO {
  IF (maxF1Score < avgOfF1[i]) THEN
    maxF1Score = avgOfF1[i];
    maxF1ScoreIndex = i;
  ENDIF
```

(continued)

Table 2 (continued)

```
    IF (maxAccScore < avgOfAcc[i]) THEN
      maxAccScore = avgOfAcc[i];
      maxAccIndex = i;
    ENDIF}

// Determine the best FET
IF (maxF1ScoreIndex == maxAccIndex) OR
(balanceDataset – False) THEN
  BFET = uniqueFETArray[maxF1ScoreIndex];
ELSE
  BFET = uniqueFETArray[maxAccIndex];
ENDIF

// Return the BFET
RETURN BFET
```

Table 3 Experimental results on the testing dataset

Vectorization approach	ML classifier	Accuracy (%)	Precision (%)	Recall (%)	F1 score (%)
Bi-grams TF-IDF	NB	85	61	76	63
	RF	97	90	79	83
	SVM	95	79	80	80
	Average	*92.33*	*76.67*	*78.33*	*75.33*
Tri-grams TF-IDF	NB	84	61	76	63
	RF	96	89	78	83
	SVM	95	79	80	79
	Average	*91.67*	*76.33*	*78*	*75*
CBoW W2V	NB	55	54	65	44
	RF	94	95	56	60
	SVM	94	47	50	48
	Average	*81*	*65.33*	*57*	*50.67*

References

1. MacAvaney, S., Yao, H.-R., Yang, E., Russell, K., Goharian, N., Frieder, O.: Hate speech detection: challenges and solutions. PLoS ONE **14**(8), e0221152 (2019). https://doi.org/10.1371/journal.pone.0221152
2. Salminen, J., Hopf, M., Chowdhury, S.A., Jung, S.G., Almerekhi, H., Jansen, B.J.: Developing an online hate classifier for multiple social media platforms. Hum. Cent. Comput. Inf. Sci. **10**, 1 (2020). https://doi.org/10.1186/s13673-019-0205-6
3. Zhang, Z., Luo, L.: Hate speech detection: a solved problem? The challenging case of long tail on Twitter (2018). http://arxiv.org/abs/1803.03662
4. Von Matthias, K.: Anti-Semitic Hate Speech in the Name of Islam. https://www.spiegel.de/international/world/wipe-out-the-jews-anti-semitic-hate-speech-in-the-name-of-islam-a-553724.html (2008)
5. Hudson, L.: Facebook Apologizes for Tolerating Violent Imagery Toward Women. https://www.wired.com/2013/01/facebook-violence-women-2/ (2013)
6. Public Media Alliance: Reporting Hate Speech, Terrorism & Violence Guidelines for Journalists and Media Workers in South East Asia. https://cdn.ethicaljournalismnetwork.org/wp-content/uploads/2019/07/FInalUNESCOSEAhatespeech.pdf (2019)
7. Robinson, D., Zhang, Z., Tepper, J.: Hate speech detection on Twitter: feature engineering v.s. feature selection. In: The Semantic Web: ESWC 2018 Satellite Events. ESWC 2018. Lecture Notes in Computer Science, vol. 11155. Springer, Cham (2018)
8. Nayel, H., Shashirekha, H.L.: DEEP at HASOC2019: a machine learning framework for hate speech and offensive language detection. In: Forum for Information Retrieval Evaluation (2019)
9. Mossie, Z., Wang, J.H.: Social network hate speech detection for Amharic language. In: Computer Science and Information Technology (2018)
10. Fatahillah, N.R., Suryati, P., Haryawan, C.: Implementation of Naive Bayes classifier algorithm on social media (Twitter) to the teaching of Indonesian hate speech. In: International Conference on Sustainable Information Engineering and Technology (2017). https://doi.org/10.1109/SIET.2017.8304122

11. Mujadia, V., Mishra, P., Sharma, D.: IIIT-Hyderabad at HASOC 2019: Hate Speech Detection. https://www.google.com/url?sa=t&rct=j&q=&esrc=s&source=web&cd=&ved=2ahUKEwjBqerE-77qAhXblXIEHS6qBlcQFjAAegQIBhAB&url=http%3A%2F%2Fceur-ws.org%2FVol-2517%2FT3-12.pdf&usg=AOvVaw3SuRvI7A3cWvyuOVAue_A9 (2019)
12. Fauzi, M.A.: Random forest approach for sentiment analysis in Indonesian language. Indones. J. Electr. Eng. Comput. Sci. **12**, 46–50 (2018). https://doi.org/10.11591/ijeecs.v12.i1.pp46-50
13. Gaydhani, A., Doma, V., Kendre, S., Bhagwat, L.: Detecting hate speech and offensive language on Twitter using machine learning: an N-gram and TFIDF based approach (2018). http://arxiv.org/abs/1809.08651
14. Ruwandika, N.D.T., Weerasinghe, A.R.: Identification of hate speech in social media. In: The International Conference on Advances in ICT for Emerging Regions, Colombo, Sri Lanka. https://doi.org/10.1109/ICTER.2018.8615517 (2018)
15. Quora Inc.: What Are Advantages & Disadvantages of TF-IDF? https://www.quora.com/What-are-the-advantages-and-disadvantages-of-TF-IDF (2015)
16. Waseem, Z.: A Comparative Study of Embeddings Methods for Hate Speech Detection from Tweets. Semantic Scholar. https://www.semanticscholar.org/paper/A-Comparative-Study-of-Embeddings-Methods-for-Hate-Waseem/043eae47d6def331ea39e41919f1500647dac44a (2018)
17. Zaghi, C.: Automatic Detection of Hate Speech in Social Media. University of Malta. https://www.google.com/url?sa=t&rct=j&q=&esrc=s&source=web&cd=&ved=2ahUKEwi89pOa8b7qAhV7mHIEHUhRBTMQFjAAegQIARAB&url=https%3A%2F%2Flct-master.org%2Fgetfile.php%3Fid%3D4125%26n%3D1%26dt%3DTH%26ft%3Dpdf%26type%3DTH&usg=AOvVaw14Gl_dt2mJ3p1EhftBscaG (2018)
18. Stack Exchange Inc.: Word2Vec: CBOW & skip-gram performance wrt training dataset size. https://stackoverflow.com/questions/39224236/word2vec-cbow-skip-gram-performance-wrt-training-dataset-size (2016)
19. Analytics Vidhya: Twitter Sentiment Analysis. https://datahack.analyticsvidhya.com/contest/practice-problem-twitter-sentiment-analysis/ (2020)

Sentiment Analysis of Stock Market Investors and Its Correlation with Stock Price Using Maximum Entropy

Liang Xue, Han Wang, Fengling Wang, and Huawen Ma

Abstract To study the correlation between the sentiment of the stock market investors and the stock price, this paper uses Python web crawler to crawl for 20 representative stocks on the Shanghai Stock Exchange (SSE) with two-year's data and the comment text of related stocks in Guba (a stock forum at Sina.com). Through the extraction and preprocessing on the crawled data, a classifier model using maximum entropy is built to classify stock-related comments into three sentiment labels for analysis. Eventually the sentiment index of investor is established to compare with the fluctuation of stock price. The comparison shows that there is a correlation between the sentiment index of investor and the volatility of stock price, which is beneficial to help investor monitor public opinion and adopts diverse strategies in the stock market.

Keywords Sentiment analysis · Stock · Crawler · Maximum entropy · Machine learning

L. Xue · H. Wang · F. Wang · H. Ma (✉)
Institute of Data Science, City University of Macau, Macau, China
e-mail: mkmhw@bitzh.edu.cn

H. Wang
e-mail: luckanny111@163.com

F. Wang
e-mail: wf1232983@163.com

L. Xue
Guangdong University of Education, Guangzhou, China

H. Wang
Zhuhai Institute of Advanced Technology, Chinese Academy of Sciences, Zhuhai, China

H. Ma
School of Information Technology, Beijing Institute of Technology, Zhuhai, China

R. Lee (ed.), *Computer and Information Science 2021—Summer*,
Studies in Computational Intelligence 985,
https://doi.org/10.1007/978-3-030-79474-3_3

1 Introduction

In recent years, the sentiment analysis of investors has attracted increasing attention in the fields of finance and investment. It represents the psychological expectations of the investors and affects the investment strategies that the investors may adopt in the future as well. To a certain extent, an accumulation of the sentiment of many investors can represent most of the investors' sentiment in the stock market. In social media, such as Guba, Weibo, and WeChat, there are a large quantity of stock market-related comment emerged almost every day. When the stock market fluctuates sharply, those comments in social media will increase largely. The correlation between the public views and the volatility of the stock price could be found by analyzing the sentiment of investors.

In this paper we use the approaches of python web crawler to grab the data of 20 stocks on the SSE (Shanghai Stock Exchange) and stock-related comment from Guba. Then we use Jieba for Chinese word segmentation and maximum entropy algorithm to create a machine learning model classifying the stock-related comments. Finally we create a sentiment index of stock investors and compare it with stock price in order to find out the correlation between them.

2 Related Works

To this day a lot of scholars have conducted plenty of research in the fields of sentiment analysis. The research methods are mainly divided into two kinds, one is using sentiment (lexicon) dictionary, and other is using machine learning.

The method based on sentiment dictionary refers to compare the word in a sentence with a word list which has value in the dictionary. Zhang et al. [1] presented a sentiment analysis method for Chinese micro-blog text based on the sentiment dictionary. Li and Shah [2] presented a new approach to learn word polarity by predicting message sentiment, using a neural network. Dhaoui et al. [3] compared the lexicon-based and machine learning approaches, showing that the two approaches are similar in the accuracy for sentiment analysis.

The method based on machine learning is to try counting the frequency of each word's occurrence in the document and train a sentiment classifier, such as using Naive Bayes, maximum entropy (ME), support vector machine (SVM), neural network, unsupervised learning, deep learning, and etc. Lee and Renganathan [4] proposed using maximum entropy technology for sentiment analysis of Chinese text. Renganathan and Low [5] used ME and SVM for sentiment analysis. Patel et al. [6] proposed using ME for sentiment analysis of big data. Alkubaisi et al. [7] used Naive Bayes to do sentiment analysis for Twitter posts. Kanavos et al. [8] presented sentiment analysis on Twitter to predict the stock price via deploying Spark streaming media to process data and using two n-grams strategies to improve classification accuracy. Zhang [9] used Naive Bayes, ME, and SVM algorithms to train and perform positive and negative sentiment analysis on the Twitter corpus. Xie et al. [10] proposed novel ME PLSA model to improve the accuracy of sentiment classification for massive data and irregular data. Cakra and Trisedya [11] used Naïve Bayes and Random Forest to classify tweets for sentiment analysis to predict the company stock price. Mohammad et al. [12] created two state-of-the-art SVM classifiers, one to detect the sentiment of tweets and short message and one to detect the sentiment of a term within a message.

For deep learning method, Dong [13] made an LSTM model to classify the sentiment of user comment and calculate the daily sentiment index of investor. Sohangir et al. [14] applied LSTM, doc2vec, and convolutional neural networks models to predict the sentiment of authors in StockTwits dataset. Yildirim et al. [15] carried out sentiment analysis of StockTwits using various deep learning approaches such as LSTM to evaluate performance. Yang et al. [16] created a sentiment analysis model based on the advantages of sentiment lexicon, word vectors, CNN, GRU and the attention mechanism to do experiment on the book review dataset.

For unsupervised learning method, Zhao et al. [17] proposed using Latent Dirichlet allocation algorithm, with financial dictionary and user group model to predict stock price by sentiment analysis on Weibo. Renault [18] found that more complex machine learning methods do not improve the accuracy of the classification, and the preprocessing method and the dataset size affect the correlation between investor sentiment and stock returns. Boiy and Moens [19] presented machine learning experiments trained from a set of example sentences which is manually annotated as positive, negative or neutral.

3 Overview of the Research

The overview of all of the research steps in this paper is shown in Fig. 1.

4 Crawling for Stock Data and User Comment

In our experiment, Python libraries such as Re, Requests, and Beautifulsoup4 are used to collect and parse stocks data through the API (http://72.push2his.eastmoney.com/api/qt/stock/kline/get). The twenty representative stocks selected on SSE are China Merchants Securities, Qingdao Haier, Bank of Communications, PingAn Bank, and Guizhou Moutai, etc., including their two-year's data (from January 31, 2017 to January 31, 2019).

The method of crawling for user comment from Guba is similar to fetching the data of stocks. Our experiment uses Python Requests library to grab data and the BeautifulSoup4 library to extract and parse the user comments. To boost the crawling speed, Python multiprocessing library can be used to crawl for the data of multiple web page simultaneously. Table 1 is the data of China Merchants Securities crawled

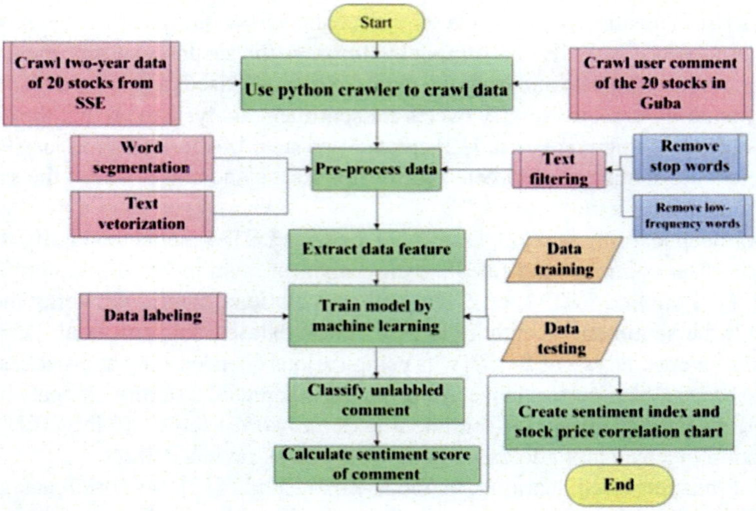

Fig. 1 Overview of the research

Table 1 The crawled data of China Merchants Securities

China Merchants Securities (600999)						
Date	Open	High	Low	Close	Volume	Amount
2017/2/3	16.76	16.85	16.47	16.53	6,953,676	115,305,048
2017/2/6	16.5	16.58	16.43	16.53	6,206,484	102,359,824
2017/2/7	16.49	16.52	16.38	16.46	7,693,441	126,504,064
2017/2/8	16.45	16.82	16.42	16.81	15,208,382	253,166,384
2017/2/9	16.75	16.92	16.7	16.82	14,237,792	239,605,968
2017/2/10	16.8	16.98	16.75	16.88	15,821,478	267,170,464
2017/2/13	16.87	17.09	16.81	16.98	17,244,438	293,200,128
2017/2/14	17	17.04	16.83	16.89	10,049,307	170,173,888
2017/2/15	17.07	17.08	16.75	16.82	14,615,835	247,842,448
2017/2/16	16.76	17.08	16.7	16.95	11,513,031	194,850,624
2017/2/17	17.03	17.44	16.94	16.96	26,267,163	452,212,544
2017/2/20	16.95	17.15	16.87	17.11	15,166,631	257,895,776
2017/2/21	17.13	17.32	17.09	17.15	16,541,192	284,444,608
2017/2/22	17.16	17.2	16.95	17.06	12,097,445	206,124,784
2017/2/23	17.1	17.11	16.83	16.89	16,499,246	278,872,640

from Easymoney.com, and Table 2 is the data of user comment crawled from Guba. All data is saved in CSV format.

5 Data Prepossessing

The preprocessing for user comment of Guba involves word segmentation, text filtering, text extraction and vectorization.

5.1 Chinese Word Segmentation

Unlike English that uses space to segment word, Chinese needs to analyze the sentence to find out the features and feature values of the words for word segmentation. Recently, the algorithms based on statistics have gradually dominated the Chinese word segmentation, such as statistical learning methods-Hidden Markov Model (HMM). HMM is to divide the words of the sentence, and calculate the probability of the result to obtain the word segmentation with the highest probability.

Based on the Viterbi algorithm of HMM, Jieba Chinese word segmentation library is used to make each word become a feature of text. The Jieba tokenizer has three kinds of tokenization [20]: precise mode, full mode, and search engine mode. We use precise mode to cut the sentence into the most precise way. Since there are many specific words in the stock market, some of them may not be recognized if we only use the default dictionary of Jieba. The solution is to add a custom dictionary into Jieba to improve the efficiency of words recognition. It is suggested that the dictionary should involve the extracted information (keywords) of the user comments and some financial thesaurus, i.e. the financial dictionary of Sogou input method.

5.2 Text Filtering: Removing Stop Words and Low-Frequency Words

The removal of stop words refers to filtering special characters in the text, and words that appear more frequently but are meaningless to determine the sentiment tendency of the text, such as Chinese character "的", "啊" and other modal auxiliary words. Those modal words usually occupy a large proportion in text vectorization, which interferes with the sentiment analysis and reduce the accuracy of sentiment classification. Therefore they must be removed first by building a stop word list or using a stop word dictionary like "HIT Stop Word List". Through defining a filter function, the stop words will be removed.

Table 2 The crawled user comment from Guba

ID	Title	Title translated in English	Date
0	换手率高, 太好了!	High turnover rate, great!	2017-03-31 05:25:00
1	招商证券 (600999) 融资融券信息 (03-30)	CMS (China Merchants Securities) (600999) Margin and securities lending information (03-30)	2017-03-31 08:07:00
2	还有那个比你们更可耻的了, 十一天十阴。	There is something more shameful, price is down ten days out of eleven days	2017-03-31 13:35:00
3	爱跑的就跑, 不用吓爷爷, 爷爷暂时不想卖。	Sell if you like, don't scare, I don't want to sell it now	2017-03-31 14:45:00
4	买了就跌 招商证券我来了 [加油]	If you buy it, it will down. I'm here at CMS. [Come on]	2017-03-31 19:32:00
5	招商证券 (600999) 融资融券信息 (03-31)	CMS (600999) Margin and securities lending information (03-31)	2017-04-01 07:42:00
6	招商证券: 各项业务保持赶超势头	CMS: Various businesses maintain momentum of catching up	2017-04-01 14:27:00
7	港口股的集体暴涨飙涨, 大幅的透支了一带一路的短期持有	The collective skyrocketing of port stocks greatly overdrafts the short-term holdings of "One Belt One Road" Initiative	2017-04-02 18:14:00
8	牛股来了 本股的技术分析和走势 必看! 会打开你思路	Bull stocks are coming. The technical analysis and trend of this stock must be seen! Will open your mind	2017-04-04 21:31:00
9	送给为定增而买股的股民: 定增原来是这样玩的	For investors who buy stocks for SPO (Secondary Public Offering): SPO is originally played like this	2017-04-05 00:29:00
10	招商证券一季度调研近200次36家券商"捧场"	CMS was surveyed by 36 brokers nearly 200 times in the first quarter	2017-04-05 10:26:00
11	券商第一获益者, 自营盘雄县导致增量资金A股成交量迅速放大	CMS in A-share market is the first beneficiary, self-operating in Panxiong County led to a rapid increase in the trading volume	2017-04-05 11:34:00
12	卖卖卖卖卖卖卖卖卖暴跌快跑卖卖卖卖坚决	Sell, sell, sell	2017-04-05 13:46:00
13	幸亏抛了, 回来看看兄弟们, 加油!	Fortunately sold off, come back to see my brothers, come on!	2017-04-05 16:59:00

(continued)

Table 2 (continued)

ID	Title	Title translated in English	Date
14	明天暴跌。。。。。。	Tumble tomorrow……	2017-04-05 18:15:00

Note There are four columns in Table 2: ID, Title, Title translated in English, and Date. The "ID" column is the number of user comment. The "Title" column is the user comment of the stock, and it is translated in the "Title translated in English" column. The "Date" column represents the posted date of comment

Fig. 2 The diagram of word cloud

After the stop words are eliminated, word frequency can be counted by using Python word cloud, collections and Re regular expression libraries. Those meaningless low-frequency words, such as words that only appear once with no obvious emotional tendency, should be removed. Some of the meaningful words with high frequency can be drawn into a word cloud diagram as shown in Fig. 2. For example, the words like "上涨 (up)", "下跌 (down)", "卖出 (selling)" have obvious emotional tendencies, while other high-frequency words like "股价 (stock price)", "成交 (trading)", and "差价 (price difference)" has no apparent emotional tendency.

5.3 *Keyword Extraction Algorithm: TF-IDF*

After word segmentation and text filtering, the next is to use the keyword extraction algorithm TF-IDF (Term Frequency-Inversed Document Frequency). Term Frequency (TF) refers to the frequency with which a given word appears in the document. The more the word appears in the document, the stronger the ability to express the document. For the word t_i in a document, it can be expressed as (1).

$$tf_{i,j} = \frac{n_{i,j}}{\sum_k n_{k,j}} \tag{1}$$

In (1), $n_{i,j}$ is the number of occurrences of the word in the document d_i, and the denominator $\sum_k n_{k,j}$ refers to the sum of the number of occurrences of all words in the document d_j.

IDF (Inverse Document Frequency) is to count the occurrences of a word in a set of documents. If a word appears in fewer documents, its ability to distinguish between documents is stronger. The IDF of a word can be obtained by dividing the total number of documents by the number of documents in which the word appears, and then taking the logarithm to the base 10 as shown in (2).

$$idf_i = lg\frac{|D|}{|\{j:t_i \in d_j\}|} \tag{2}$$

$|D|$ represents the total number of documents in the corpus. $|\{j:t_i \in d_j\}|$ indicates the number of files containing the word t_i (i.e. n_i means the number of document when $j \neq 0$). The smaller the value of $|\{j:t_i \in d_j\}|$, the larger the value of idf tends to be. In order to avoid the situation where the new words have not appeared in the corpus and the denominator is zero, the denominator plus 1 is generally used which is called Laplace smoothing method, i.e. $1 + |\{j:t_i \in d_j\}|$ [21]. Multiplying tf and idf gets the formula $\text{tf-idf}_{i,j} = \text{tf}_{i,j} * \text{idf}_i$. When the frequency of occurrence of a word is high in a document and it is low in all documents, the value of tf-idf is high. Hence the word with the highest tf-idf value is the most suitable keyword for the document.

5.4 Text Vectorization and Word Matrix

The prerequisite of text classification is to vectorize the text and extract its features, that is, to express the text as a computer-recognizable real number vector. The bag-of-words model means when a word in a single text appears in the dictionary, its vector value is increased by 1, and the value is equal to the total number of word occurrences. As a weighted method, TF-IDF keyword extraction algorithm assigns weights to the frequency of words appearing on the basis of the bag-of-words model.

For text feature extraction, the two most important steps are to use the TF-IDF vector generator and convert the original document into a word-document matrix. The larger the data set, the more advantageous the use of TF-IDF Vectorizer for text vectorization [22]. The word-document matrix stores a sparse matrix. The columns of the matrix represent the features whilst the rows represent the number of original documents. The value of the matrix represents the TD-IDF value of the features contained in the document, ranging from 0 to 1.

6 Sentiment Classification of User Comment

In natural language processing, there exists two kinds of approaches for sentiment analysis of text. One is using sentiment dictionary and the other is machine learning method.

Comparing the words in a sentence with sentiment dictionary one by one, we will get the sentiment score of each word. Then performing a weighted calculation to sum all the emotionally inclined words in the sentence, we will get all of the sentiment score of the entire sentence gradually.

The method of machine learning requires selecting a certain number of comment texts that is representative, doing segmentation and labeling the text manually with a positive or negative tag. Then those labeled data is divided into the training set and the test set for training the model. Once the model has been built, other unlabeled comment texts are able to be classified and labeled by the model. Figure 3 is the processes to do text classification by machine learning method of building a Maximum Entropy (ME) model.

6.1 *Building a Classification Model of ME*

The following is the process of building a classification model of ME:

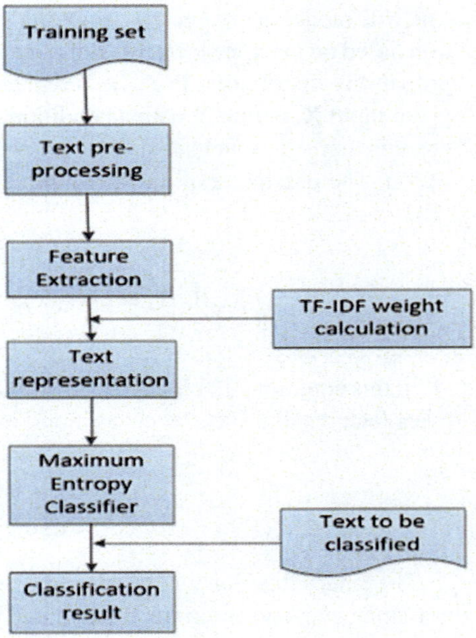

Fig. 3 The processes of ME classification of machine learning

6.1.1 Calculation of Entropy

Entropy was originally introduced in thermodynamics to evaluate the uncertainty of the system. Shannon introduced entropy to quantify the expected value of the information. The more chaotic the message is, the higher the entropy is. N-gram is one of the most traditional and popular structures used in natural language and text mining areas. It is a probabilistic language model based on (n − 1) order Markov chains. It refers to n words that appear continuously in the text, and the structure of the sentence is inferred by the probability of n words [23]. In order to improve the accuracy of classification, n-grams which cannot express any emotion will be discarded [8]. The n-gram probability distribution for calculating entropy H is shown in (3).

$$\mathrm{H(g)} = -\sum_{i=1}^{n} P(x_i)\log P(x_i) \tag{3}$$

In (3), g is a specific n-gram, $x_1 \ldots x_n$ is the possible result, which is represented by the probability $P(x_1) \ldots P(x_n)$. $\sum$ represents the sum of possible values of the variable, and n is the number of emotions (when $n = 3$, it means that positive, negative and neutral emotions are included).

6.1.2 Building Processes of ME Model

The ME model is obtained by applying statistical learning principles for classification based on machine learning. Suppose that the classification model is a conditional probability distribution P(Y|X), X represents input, and Y represents output. For a given input X, output Y with a conditional probability P(Y|X). The principle of ME ensures that the model satisfies all known constraints and chooses the best classification. The parameter of the maximum entropy estimation of P(c|f) is shown in (4) [24].

$$\mathrm{P(c|f)} = \frac{1}{\mathrm{Z}(f)}\exp\left(\sum_{i}\lambda_{i,c}F_{i,c}(f,c)\right) \tag{4}$$

In this equation, $Z(f)$ is a normalization function, and $F_{i,c}$ is a binary function that takes features and class labels as input. It is defined as (5).

$$F_{i,c}(f,c') = \begin{cases} 1, \ n(f) > 0 \ and \ c' = c \\ 0, \ otherwise \end{cases} \tag{5}$$

The binary function will be triggered when there are certain features (units, binary grammar, etc.) and emotions are assumed in some way. Similarly, no assumptions are made about the conditional relationship between each element. λ is the vector of the weighted parameter of the feature set. Those parameter variables are estimated by the technique called MegaM created by Daumé [25]. MegaM iteratively updates the values of the parameters that satisfy the constraints of the comments while continue to maximize the entropy of the model, and iteratively converges the model to the best maximum value of the probability distribution.

When fetching the data of user comment is completed, we can randomly select 1000–2000 pieces of comment and manually labeled with positive, neural, and negative for sentiment classification. The label "1" means positive, "−1" means negative, and "0" means neutral. Then, divide the dataset by 60% for training set and 40% for test set, using cross-validation method for the training of ME model. When the ME model classifier has been trained, it can be used to classify other unlabeled comment

Table 3 Labeling on the comment text of China Merchants Securities

Title	Title translated in English	Date	sentiment_tag
对我们股民应该有一个交代!	There should be an explanation to our investors!	2017-04-05 21:08:00	−1
小散都跑完了	All the individual investors has sold off	2017-04-05 21:08:00	−1
你可以不信	You can't believe it	2017-04-05 21:09:00	0
上午好好的捨不得出	Can't give up in the morning	2017-04-05 21:09:00	1
又跳水了	Diving again	2017-04-05 21:09:00	−1
白写了那么多	Wasted so much writing	2017-04-05 21:10:00	−1
看来是没救了	Seems hopeless	2017-04-05 21:10:00	−1
可以加仓抄底了	You can buy more in the bottom	2017-04-05 21:11:00	1
快跑快跑	Sell off immediately	2017-04-05 21:11:00	−1
节前多说两句	Say a few more words before the holiday	2017-04-05 21:11:00	0
可以加仓抄底了	You can buy more in the bottom	2017-04-05 21:12:00	1
没有托	It's real	2017-04-05 21:12:00	0
股票有点贵	The stock is a bit expensive	2017-04-05 21:13:00	−1
毫无活力	No vitality	2017-04-05 21:14:00	−1

Note The value of "sentiment_tag" column represents the types of sentiments. When its value is −1, it means the negative sentiment, and value 0 means neutral and value 1 means positive. The "Title" column is the comment of users and it is translated in English as the next column "Title translated in English". The "Date" column means the posted date of comment

Table 4 The performance of ME model

Performance	ME model
Accuracy	0.723
Precision	0.762
Recall	0.421

texts. Finally all comment texts are labeled. Table 3 is an example of labeling on the comment text of users.

To measure the ME model performance, precision, recall and accuracy can be calculated through Formulas (6), (7) and (8) respectively, as listed below:

$$\text{Precision} = \frac{\text{TP}}{\text{TP} + \text{FP}} \tag{6}$$

$$\text{Recall} = \frac{\text{TP}}{(\text{TP} + \text{FN})} \tag{7}$$

$$\text{Accuracy} = \frac{(\text{TP} + \text{TN})}{(\text{TP} + \text{FP} + \text{FN} + \text{TN})} \tag{8}$$

TP: True Positive, FP: False Positive, TN: True Negative, FN: False Negative.

The evaluation result is listed in Table 4.

6.2 *Calculation of the Sentiment Index of Investor*

After user comments are categorized into different sentiment labels, the number of positive, negative, and neutral comments on each trading day can be counted. Referring to the method proposed by Frank and Antweiler to construct investor sentiment index based on the comments of stock forum, we can define (9) to calculate the index of investor sentiment.

$$\text{Sentiment} = \frac{\text{Pos} - \text{Neg}}{Pos + Neg} \tag{9}$$

In (9), *Pos* means the total number of positive comment posts of the investors, while *Neg* means the total number of negative posts. Sentiment represents the average score of the investor sentiment, ranging between -1 and 1. Since the sentiment value of neutral posts is zero, it has not been added into the formula during the calculation in terms to exclude the influence of meaningless text like advertisement.

As an example shown in Table 5, the total number of comment posts is 31 on April 5th 2017, of which 13 are bullish (positive), and 18 are bearish (negative), so the sentiment index is: $\frac{13-18}{31} = -0.1613$. On April 11th the total number of posts is

Table 5 The calculation example of the investor sentiment index of China Merchants Securities

Date	Price	Posts	Bullish	Bearish	Sentiment index
2017/4/5	16.47	31	13	18	−0.1613
2017/4/6	16.48	19	14	5	0.4737
2017/4/7	16.47	4	1	3	−0.5000
2017/4/8	16.47	13	8	5	0.2308
2017/4/10	16.48	5	3	2	0.2000
2017/4/11	16.45	12	7	5	0.1667
2017/4/12	16.57	9	6	3	0.3333
2017/4/13	16.46	15	9	6	0.2000

12, of which 7 are bullish and 5 are bearish, so the sentiment index is: $\frac{7-5}{12} = 0.1667$. As seen from Table 5, the sentiment index fluctuates between -1 and 1.

7 Analysis of Experiment Result

The analysis of the experiment result has two aspects, one is the analysis of the investor sentiment index and the other is the comparison of the sentiment index of investor and the stock price.

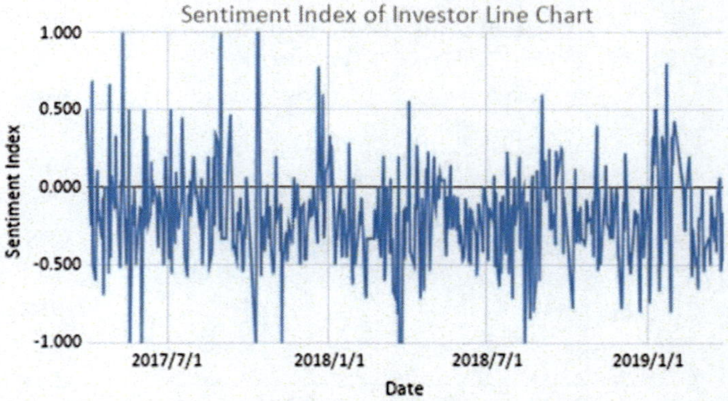

Fig. 4 The sentiment index of investor line chart

7.1 *Daily Analysis for the Sentiment Index of Investor*

As shown in Fig. 4, from January 31, 2017 to January 31, 2019, except Saturdays, Sundays and holidays, the fluctuation of the sentiment index of investor of China Merchants Securities in 2017, is greater than in 2018. In May, August, and October 2017, the sentiment index of investor reached a relative peak, and then reached the bottom in May, June, and October 2017. From the second half of 2018 to January 2019, it was more stable than in 2017.

7.2 *Comparative Analysis of Sentiment Index of Investor and Stock Price*

Compared the sentiment index of investor with the price of China Merchants Securities during the trading days from 2017 to 2019, the investor sentiment index is somewhat consistent with the fluctuation of stock price, as shown in Fig. 5. The red line represents the stock price while the blue line represents the sentiment index. During the rapid rising of the stock price in August and October 2017, the sentiment index of investor was also up and then reached its peak. From August to December 2018, the stock price was down slightly and then turned into the sideways period, and the sentiment index of investor was down as well.

Fig. 5 The comparison of the sentiment index of investor and the stock price

8 Conclusions and Future Work

After crawling for the two-year data of 20 stocks on the SSE and the stock-related user comment in Guba, we pro-process the Chinese text and build a text classification model of maximum entropy, and then calculate the sentiment index of investor, comparing with the trends of stock price to find out the correlation between them. The experiment result shows that there is a somewhat correlation between the index of investor sentiment and the fluctuation of stock price. It would be helpful for investors to monitor public opinion in social media and adopt appropriate strategy through analyzing the correlation between the sentiment index and the trend of stock price.

For future work, the first is to consider using different machine learning or deep learning algorithms to build models, compare the performance of different models, and figure out the algorithm which has better performance. The second is to improve the sentiment dictionary, especially increasing the terminologies in the fields of finance and investment, so as to make the word segmentation more accurate and enhance the classification accuracy of the model.

References

1. Zhang, S., Wei, Z., Wang, Y., Liao, T.: Sentiment analysis of Chinese micro-blog text based on extended sentiment dictionary. Future Gener. Comput. Syst. **81**, 395–403 (2018)
2. Li, Q., Shah, S.: Learning stock market sentiment lexicon and sentiment-oriented word vector from stocktwits. In: Proceedings of the 21st Conference on Computational Natural Language Learning (CoNLL 2017), pp. 301–310 (2017)
3. Dhaoui, C., Webster, C.M., Tan, L.P.: Social media sentiment analysis: lexicon versus machine learning. J. Consum. Mark. **34**(6), 480–488 (2017)
4. Lee, H.Y., Renganathan, H.: Chinese sentiment analysis using maximum entropy. In: Proceedings of the Workshop on Sentiment Analysis Where AI Meets Psychology (SAAIP 2011), pp. 89–93 (2011)
5. Renganathan, H., Low, B.W.: Sentiment analysis using maximum entropy and support vector machine. In: Knowledge Engineering Conference (STAKE 2010), p. 1 (2010)
6. Patel, D., Saxena, S., Verma, T.: Sentiment analysis using maximum entropy algorithm in big data. Int. J. Innov. Res. Sci. Eng. Technol. **5**, 8356–8361 (2016)
7. Alkubaisi, G.A.A.J., Kamaruddin, S.S., Husni, H.: Stock market classification model using sentiment analysis on Twitter based on hybrid Naive Bayes classifiers. Comput. Inf. Sci. **11**(1), 52–64 (2018)
8. Kanavos, A., Vonitsanos, G., Mohasseb, A., Mylonas, P.: An entropy-based evaluation for sentiment analysis of stock market prices using Twitter data. In: 2020 15th International Workshop on Semantic and Social Media Adaptation and Personalization (SMA), pp. 1–7. IEEE (2020)
9. Zhang, L.: Sentiment analysis on Twitter with stock price and significant keyword correlation. Doctoral dissertation (2013)
10. Xie, X., Ge, S., Hu, F., Xie, M., Jiang, N.: An improved algorithm for sentiment analysis based on maximum entropy. Soft. Comput. **23**(2), 599–611 (2019)
11. Cakra, Y.E., Trisedya, B.D.: Stock price prediction using linear regression based on sentiment analysis. In: 2015 International Conference on Advanced Computer Science and Information Systems (ICACSIS), pp. 147–154. IEEE (2015)

12. Mohammad, S.M., Kiritchenko, S., Zhu, X.: Building the state-of-the-art in sentiment analysis of tweets. arXiv preprint arXiv:1308.6242 (2013)
13. Dong, C.J.: Research on the Construction of Investor Sentiment Index in Stock Market. Henan University (2020)
14. Sohangir, S., Wang, D., Pomeranets, A., Khoshgoftaar, T.M.: Big data: deep learning for financial sentiment analysis. J. Big Data **5**(1), 3 (2018)
15. Yıldırım, S., Jothimani, D., Kavaklioğlu, C., Başar, A.: Deep learning approaches for sentiment analysis on financial microblog dataset. In: 2019 IEEE International Conference on Big Data, pp. 5581–5584 (2019)
16. Yang, L., Li, Y., Wang, J., Sherratt, R.S.: Sentiment analysis for E-commerce product reviews in Chinese based on sentiment lexicon and deep learning. IEEE Access **8**, 1 (2020)
17. Zhao, B., He, Y., Yuan, C., Huang, Y.: Stock market prediction exploiting microblog sentiment analysis. In: 2016 International Joint Conference on Neural Networks (IJCNN), pp. 4482–4488. IEEE (2016)
18. Renault, T.: Sentiment analysis and machine learning in finance: a comparison of methods and models on one million messages. Digit. Finance (2019)
19. Boiy, E., Moens, M.F.: A machine learning approach to sentiment analysis in multilingual web texts. Int. J. Inf. Retr. **12**(1), 526–558 (2009)
20. Jiang, W., Ying, C.: Sentiment Analysis of Text Based on Spark. Available from https://developer.ibm.com/zh/articles/cc-1606-spark-seniment-analysis (2016). Accessed 10 Mar 2021
21. TF-IDF: Wikipedia. Available from https://zh.wikipedia.org/wiki/Tf-idf (2020). Accessed 3 Mar 2021
22. cathy1997. Jianshu, 2019. Available from https://www.jianshu.com/p/f73a8c639cfe (2019). Accessed 3 Mar 2021
23. n-gram. Wikipedia. Available from https://zh.wikipedia.org/wiki/N%E5%85%83%E8%AF%AD%E6%B3%95 (2020). Accessed 11 Mar 2021
24. Mehra, N., Khandelwal, S., Patel, P.: Sentiment Identification Using Maximum Entropy Analysis of Movie Reviews. Stanford University, USA (2002)
25. Daumé, III, H.: Notes on CG and LM-BFGS optimization of logistic regression. 198:282 (2004). Paper available at http://pub.hal3.namedaume04cg-bfgs. Implementation available at http://hal3.name/megam

Intrusion Detection for Modern DDos Attacks Classification Based on Convolutional Neural Networks

Wenwen Chen, Haiyan Zhang, Xiaoshu Zhou, and Yangjie Weng

Abstract Distributed Denial of Service (DDos) attack is one of the most harmful attacks demonstrating its huge scale and enormous impact on people's daily life and also companies. Intrusion detection is an effective method for detecting and responding to it. But due to different characteristics brought by modern DDos attacks, traditional intrusion detection is not as useful as before. New researches on combination between deep learning algorithms and intrusion detection show high efficiency and accuracy. Thus, in this paper, we propose an intrusion detection measure to classify modern DDos attacks using Convolutional Neural Network (CNN), a deep learning algorithm. The dataset utilized for training and testing the model is conducted in 2019 and includes modern DDos attacks. The testing results show that CNN has the highest accuracy on DDos attack detection compared to other selected models, Support Vector Machine (SVM) and Random Forest. The performances of CNN with various multi-hidden layer amounts are similar.

Keywords DDos attacks · CNN · Intrusion detection

1 Introduction

With the rapid development of the Internet, it becomes an essential part of people's daily life and also companies regardless of size. Billions of sensitive data are stored on the Internet, which attracts attackers to launch attacks for tremendous profit. Denial of Service (Dos) attacks are one of the most harmful attacks on the Internet due to its impact on legitimate users. Attackers launch Dos attack mainly to disrupt availability

W. Chen (✉) · H. Zhang · X. Zhou
Department of Computer Technology, Beijing Institute of Technology, Zhuhai, No. 6 Jinfeng Road, Tangjiawan, Zhuhai, Guangdong, China
e-mail: alicecwwchen@163.com

Y. Weng
Zhuhai Central Sub-Branch, The People's Bank of China, Zhuhai, No. 63 Wenyang Road, Zhuhai, Guangdong, China
e-mail: wengyjsysu@live.cn

R. Lee (ed.), *Computer and Information Science 2021—Summer*,
Studies in Computational Intelligence 985,
https://doi.org/10.1007/978-3-030-79474-3_4

of users' vital services. In the summer of 1999, the Computer Incident Advisory Capability (CIAC) reported the first Distributed DoS (DDoS) attack incident and most of the DoS attacks since then have been distributed in nature [1].

In recent years, the scale of DDos attacks is getting bigger and the frequency is getting higher. 7.9 million DDos attacks are launched in 2018 and the amount will double to 15.4 million in 2023 according to Cisco's survey [2]. Securelist's analysis of DDos attacks on Quarter One, 2020 shows that a clear rise in both the average and maximum DDos attack duration [3]. The peak number of DDos attack per day is 242. On February 28, 2018, GitHub was hit by what may be the largest DDoS attack to date, with a peak traffic of 1.35 Tbps per second. Ten minutes after the two intermittent attacks, GitHub turned to Akamai, a CDN service provider, for help. The attack ended eight minutes after Akamai intervened. Amazon Web Services, was hit by a gigantic DDoS attack in February 2020 [4]. The attack relies on vulnerable third-party CLDAP servers and amplifies the amount of data sent to the victim's IP address by 56–70 times, which lasted for three days and peaked at an astounding 2.3 Tbps [4].

To solve the problems brought by DDos attacks, several measures are proposed by researchers. As one of the effective measures, intrusion detection shows its potential on monitoring the events occurring in a network and detecting DDos attacks. It consists of information sources, analysis and response [5]. Information sources includes where the events occur, such as hosts and network and statistical data of the events. Analysis determines which events are attacks and when the attacks take place. There are two popular method used in analysis, Machine learning and deep learning. Machine learning addresses the question of how to build computers that improve automatically through experience, which is one of today's most rapidly growing technical fields, lying at the intersection of computer science and statistics [6]. Intercross research between machine learning and other subjects are hot topics nowadays. Intrusion detection is one of it. SVM, Random forest and Naive Bayes are some of algorithms applied to it. Deep learning allows computational models that are composed of multiple processing layers to learn representations of data with multiple levels of abstraction, which is applied to improve the state-of-the-art in speech recognition, visual object recognition, object detection and many other domains such as drug discovery and genomics [7]. It is also combined with traditional intrusion detection methods to enhance the accuracy of detection. Convolutional Neural Network (CNN), RNN, LSTM and the combination of them are useful models in the researches for intrusion detection specially for DDos attacks. Response is actions taken to deal with attacks, either positive or passive. The first two steps of intrusion detection are the emphasis of this paper.

The rest of the paper is organized as following: related work discusses related work in machine learning and deep learning aspects. Methodology describes the CNN model characterizes and process of conducting this project. System Implementation shows the dataset selected for analysis and details of every step of the project. Experimental results and analysis display multiple metrics used for evaluation and comparison between different models. Conclusion gives a conclusion of the whole paper and depicts the future work.

2 Related Work

2.1 Machine Learning

In Wang's paper, a new intrusion detection approach, called FC-ANN, based on ANN and fuzzy clustering is proposed [8]. The experimental results using the KDD CUP 1999 dataset demonstrates the effectiveness of the approach especially for low-frequent attacks, i.e., R2L and U2R attacks in terms of detection precision and detection stability [8]. Kokila applies Support Vector Machine (SVM), Bagging, Radial Basis Function Network, J48 Decision Tree, Random Forest, Naive Bayes and Bagging on existing DARPA dataset to detect DDos attacks and finds out that SVM gives less false positive rate and high classification accuracy [9]. He et al. identifies DDoS attacks from the source side in the cloud using nine ML algorithms [10]. It also shows high accuracy (99.7%) and low false positives (<0.07%) for SVM [10]. Information of where and when dataset is collected is not given. Combination of different ML algorithms is also utilized by Kuang et al. [11]. A novel SVM model combining kernel principal component analysis (KPCA) with genetic algorithm (GA) is proposed for intrusion detection [11].

2.2 Deep Learning

Haider et al. [12] use a deep CNN Ensemble Framework on DDoS Attack Detection in Software Defined Network. CNN is utilized with RNN and LSTM respectively on CICIDS-2017 to identify DDos attacks from the flow data. The experimental results show the Ensemble CNN outperforms the other three proposed DL-approaches in almost all the evaluation metrics results such as accuracy [12]. A multi-channel CNN (MCNN) is proposed in Chen et al.'s paper and selected features are split into three parts, traffic, Packet and Host features [13]. They are treated as three channels in CNN. The results show that MCNN has the highest accuracy contrasted with single channel CNN, LSTM and shallow models [13]. Wu et al. compare the capability of CNN with different number of multi-hidden layers and sizes of convolution kernels to classify four network attacks on NSL KDD dataset [14]. It shows lower accuracy than RNN model but higher FAR. To preserve attack features of input traffic data, LSTM-RNN is introduced in training phase to generate classifiers in Fu et al.'s paper [15].

Although researches above found out that deep learning has an excellent performance in DDoS detection, there is still the lack of classifier for multiply types of DDos attacks with modern DDos attacks.

3 Methodology

3.1 CNN Model

CNN takes this name from mathematical linear operation between matrixes called convolution, which contains four types of layers [16]. The main idea is to compress the dimension and size of original data to converge and make final classification. It shows the potential for image data classification especially largest image data set (Image Net), computer vision and natural language processing (NLP) [16].

The DDos classifier based on CNN contains four steps displayed in Fig. 1.

3.1.1 Data Processing

Data Preprocessing aims to transform the original data into the training data. Its purpose is to obtain better training data features and make the machine learning model approach the upper limit. Since CICDDoS2019 is a well-organized dataset, part of data preprocessing has been conducted. All data except Flow ID, source IP and destination IP have been converted to numerical data. Miss values have been supplemented. For further analysis, data scaling, label encoding and feature selection are still needed in this stage.

3.1.2 Model Construction and Training (CNN Model)

CNN model treats input as images (with 2/3 dimensions) and processes the images to Convolutional layer, non-linearity layer, pooling layer and fully-connected layer in the end (Fig. 2). Through the Convolutional layer, features are abstracted to a feature map using numerical neurons. It depicts partial characteristics of the input images. In the Convolutional layer, filter or kernel number and filter or kernel depth are parameters can be adjusted. Main task of the pooling layer is to reduce dimension

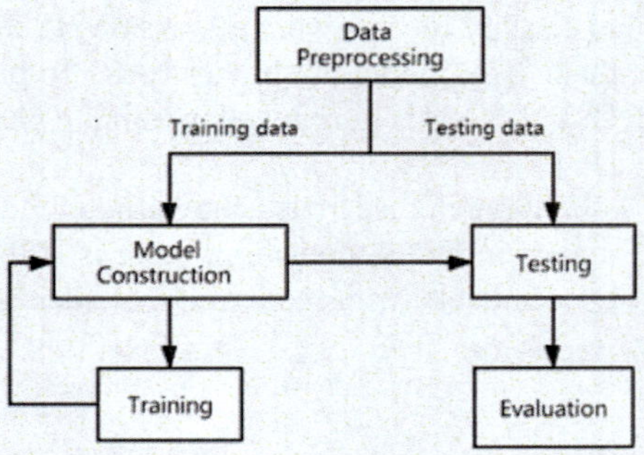

Fig. 1 Flow chart of the DDos classifier based on CNN

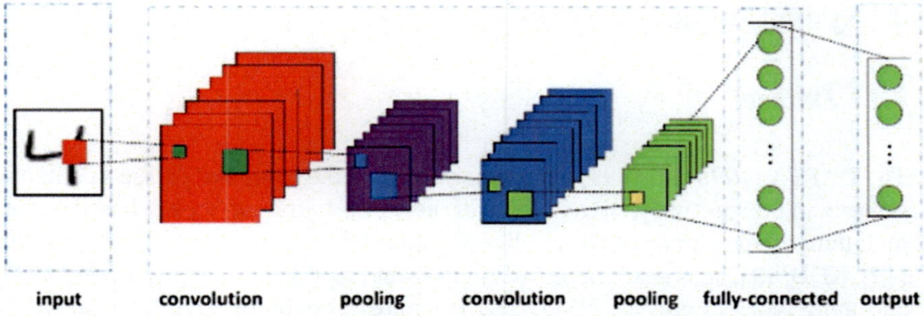

Fig. 2 The architecture of the convolution neural network introduced by Kehe Wu

of the output of last layer. It is conducted by aggregating multiple neuron outputs into one neuron in next layer. Max and average pooling are two kinds of pooling measures using maximum or average values. As the same as multi-layer perceptron neural network, each node in one layer is fully connected to nodes in the next layer, Fully connected (FC) layers. Each node in these layers is a neuron with a nonlinear activation function. Additionally, a flattened matrix needs to be conducted before FC layers. Conduction of a CNN model for DDos attacks classifier include design of these types of layers, which will be discussed in next section. After model conduction, the next step is to train the model with training dataset obtained from the last phase.

3.1.3 Testing

The testing datasets is utilized in the trained model gained after the last step to obtain metrics such as accuracy. If the metrics don't meet the expectation, the model will be redesigned by adjusting parameters of different layers and trained again. Through this iteration, a better performance can be got.

3.1.4 Evaluation

Evaluation of performance of the trained model is conducted by calculating metrics of the results gained in the last phase.

4 System Implementation

4.1 Dataset

The DARPA (1998 and 1999) data set is one of the most popular datasets containing seven and, respectively, five weeks of network traffic in packet-based format, including various kinds of attacks like DoS, buffer overflow, port scans, or rootkits [17]. KDD'99 Cup is based on DARPA (1998 and 1999) data with more features (41) and more types of attacks. NSL-KDD is another version of KDD'99 Cup without redundant records. DoS, probe, U2R, R2L are included in the dataset. The DARPA, KDD'99 and NSL-KDD datasets are packet-based data, providing payload or statistical features for further analysis in packet level. But the focus of our research is flow-based. CICIDS-2017 dataset is derived from a simulative business environment comprised of at least 80 features of network traffic. It contains flow-based data and classifies numerous attacks including botnet, cross-site-scripting, DoS and so on. The DARPA, KDD'99, NSL-KDD and CICIDS-2017 datasets are popular data focusing on multiple types of attacks. DDos attack is not the primary concern, which is not suitable for our research.

DDoS 2016 is a dataset focusing on only four kinds of DDoS attacks, SIDDOS, smurf, UDP flood, and HTTP flood. The emulated network environment information is not contained in the dataset. CICDDoS2019 is a dataset collected and reorganized to flow-based data with well-defined features and labels recently in 2019.

Compared to DDoS 2016, it captured more types of DDos attacks (13) including modern reflective DDos attack such as NTP, NetBIOS, SSDP, UDP-Lag, and TFTP. CICDDoS2019 divides attacks into two categories, reflective attacks (10) and exploitation attacks (3) shown in Fig. 3. The testbed of CICDDoS2019 consists of two completely separated networks, namely Attack-Network and Victim Network [17].

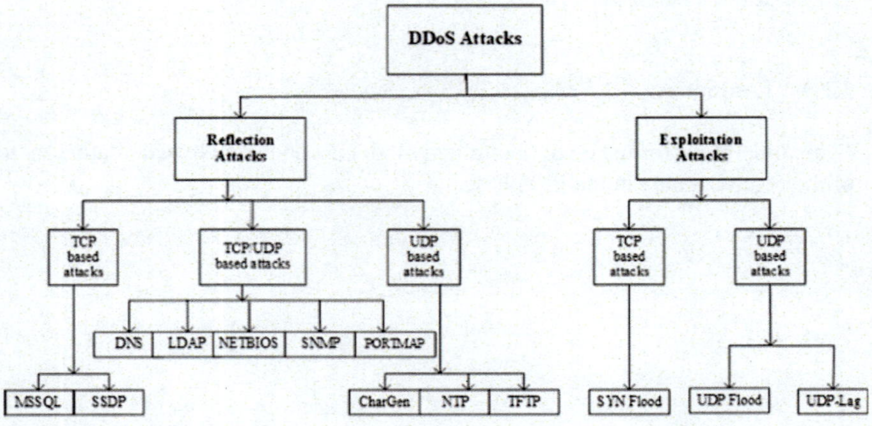

Fig. 3 DDoS attack taxonomy introduced by Iman Sharafaldin

Table 1 Distributions of different data types

	Training data	Testing data	Total
MSSQL	2503	33	2536
NETBIOS	19,979	4969	24,948
NTP	20,000	5000	25,000
SYN	19,586	5000	24,586
UDP-Lag	19,994	4998	24,992
Benign	17,669	5000	22,669
Total	99,731	25,000	124,731

All commonly used and necessary equipment including router, firewall, switch, along with the different versions of the commonly used operating systems are employed in the testbed [17]. The B-Profile for the dataset extracts the abstract behavior of 25 users based on the HTTP, HTTPS, FTP, SSH, and email protocols [17]. It shows that CICDDoS2019 is conducted in an experimental environment closer to real network in contrast to other datasets mentioned above. Additionally, more than 80 features are included in the dataset. Based on advantages of CICDDoS2019 and the goal to focus on DDos attack and conduct classification, we choose it as the information resource.

Based on Fig. 3, there are five categories of DDos attacks, TCP-based reflection attack, TCP/UDP-based reflection attack, UDP-based reflection attack, TCP-based exploitation attack and UDP-based exploitation attack. Five DDos attacks from these categories are selected for analysis including MSSQL, NETBIOS, NTP, SYN and UDP Lag attacks.

4.2 Dataset Distribution

Total amount of samples is 124,731. There are 99,731 samples for training and 25,000 samples for testing. Five types of DDos attacks have close number of samples in either train dataset or test dataset, which is close to 20,000 samples and 5000 samples respectively in Table 1. Some benign data is also contained in the dataset with 2503 training samples and 33 test samples.

4.3 Data Preprocessing

4.3.1 Feature Selection

After investigating the whole dataset, we find a feature named "Unname:0" contain unique values like ID for each flow data. It is then removed from all features. Then,

Table 2 Correlation matrix of training dataset

Feature	Feature				
	Source port	Destination port	…	Inbound	Label
Source port	1.00	0.06	…	−0.16	0.02
Destination port	0.06	1.00	…	0.20	0.05
…	…	…	…	…	…
Inbound	−0.16	0.20	…	1.00	0.22
Label	0.02	0.05	…	0.22	1.00

feature selection is conduct by comparing the correlation of different features with attacks. Correlation matrix is used in this step. A correlation matrix depicts the correlation coefficients between features in a table. Each cell in the table displays the correlation between two features. The higher absolute value of each cell, the closer correlation between two features. In order to gain it, the split of training and testing datasets is conducted first. 80% of dataset is allocated to training dataset and the rest is testing dataset. The correlation matrix of training dataset is produced using Pandas library in python and shown in Table 2. Top 24 features with the highest correlation with attack type are selected and displayed in Table 3.

4.3.2 Data Scaling

Statistical data of four example features including mean, standard deviation, maximum value and minimum value are shown in Table 4. It is displayed that different types of measuring units are used for different features including bytes, byte/s, seconds and so on. Some features are different statistical data like mean or standard deviation. It is obvious that there is numerical variation among different feature data, which will bring negative impact on training results.

Thus, it is necessary to process data scaling. The main data scaling method is normalization, which shifts data with different units to a certain range and under the same order of magnitude. Z-score normalization is chosen for data scaling before next step, constructing CNN models. In this method, the mean and standard deviation of the original data are given to standardize the data. After processing, the feature data conform to the standard normal distribution. The mean of the data is 0 and the standard deviation is 1. As a result, different distributions effect of various features will be reduced. μ represents mean and σ represents standard deviation.

$$Z = \frac{x - \mu}{\sigma} \tag{1}$$

After Z-score normalization, four selected feature data are re-scaled and shown in Table 5.

Table 3 Selected features

Features	Correlation
Protocol	−0.75535
Fwd packet length max	−0.6784
Packet length mean	−0.67823
Avg fwd segment size	−0.67104
Fwd packet length mean	−0.67104
Max packet length	−0.65979
Min packet length	−0.64857
Average packet size	−0.64688
Fwd packet length min	−0.645
Fwd packets/s	−0.24106
Idle std	0.419965
Flow duration	0.436695
Fwd IAT total	0.437258
Init_Win_bytes_forward	0.495542
Idle min	0.517399
Flow IAT max	0.530127
Idle max	0.530575
Fwd IAT max	0.53083
Idle mean	0.545993
Fwd IAT std	0.57157
Flow IAT std	0.574075
Flow IAT mean	0.580912
Fwd IAT mean	0.586028
ACK flag count	0.803378

Table 4 Statistical data of four example features before scaling

	Protocol	Fwd packet length max	Fwd IAT mean	ACK flag count
Mean	12.77	266.81	1,581,831.00	0.37
Std	5.36	275.01	3,140,355.00	0.48
Min	0.00	0.00	0.00	0.00
Max	17.00	2935.00	30,890,630	1.00

4.3.3 Label Encoding

This is a multi-classes classification, label encoding measure is needed to be conducted to distinct different attacks. Since the types of DDos attack are discrete data, one-hot encoding is utilized in this step. If the total amount of flow data types

Table 5 Statistical data of four example features after scaling

	Protocol	Fwd packet length max	Fwd IAT mean	ACK flag count
Mean	0.00	0.00	0.00	0.00
Std	1.00	1.00	1.00	1.00
Min	−2.39	−0.97	−0.51	−0.77
Max	0.79	9.74	9.35	1.30

is N, we set a vector of N dimension is for identification. We convert Benign flow data to [0, 0, 0, 0, 0, 1], MSSQL attacks to [0, 0, 0, 0, 1, 0], NETBIOS attacks to [0, 0, 0,1, 0, 0], NTP attacks to [0, 0, 1, 0, 0, 0], SYN flow to [0, 1, 0, 0, 0, 0] and UDP Lag to [1, 0, 0, 0, 0, 0].

4.3.4 Model Construction, Training and Testing

From Wu et al.'s experimental result [14], CNN models with various amount of multi-hidden layers (combination of Convolutional layer and pooling layer) show different performances. Thus, CNN models with distinct multi-hidden layers are trained and tested in this step. Table 6 indicates parameters including the kernel size and stride of Convolutional layer and pooling layer set for CNN model with one multi-hidden layer and two multi-hidden layers. CNN-1 represents CNN model with one multi-hidden layer and CNN-2 represents CNN model with two multi-hidden layers. In order to solve the overfitting problems of the algorithm caused by incremental network layers, ReLU function and Dropout are used in the model. The CNN-1 and CNN-2 model are displayed in Figs. 4 and 5.

Table 6 Parameters of CNN models

Parameter	Model	
	CNN-1	CNN-2
Convolution layer1-kernel size	3 * 6	2 * 6
Convolution layer1-strides	1	1
Convolution layer1-filter	32	32
Pooling layer1-kernel size	4 * 1	2 * 5
Pooling layer1-strides	1	1
Convolution layer2-kernel size	–	2 * 2
Convolution layer2-strides	–	1
Convolution layer2-filter	–	64
Pooling layer1-kernel	–	1 * 1
Pooling layer2-strides	–	1

Fig. 4 CNN-1 model

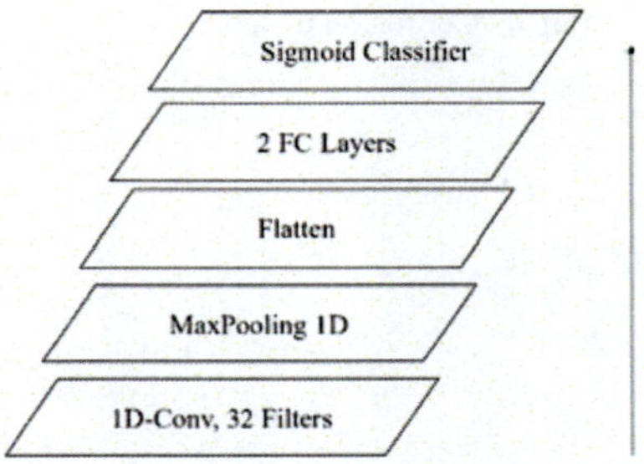

Fig. 5 CNN-2 model

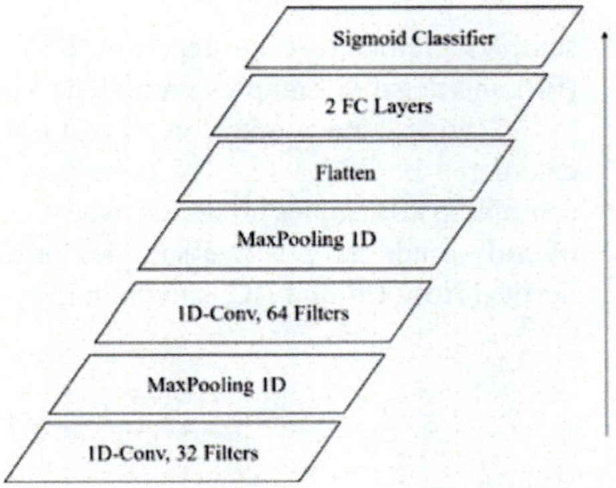

5 Experimental Results and Analysis

As shown in Table 7, the experiment is set up on a DELL computer with Core i7 CPU and 6G memory. The operating system is Microsoft Windows 10. Tensorflow 2.4 is chosen for CNN model construction, training and testing. Scikit Learn, Pandas are selected libraries implemented in the data preprocessing, including data scaling, label coding and features selection.

The performance of the CNN model is evaluated based on different metrics generated from testing results. They contain accuracy (AC), precision (PR), recall (RC), F1 score and testing time. Before giving the definition of these metrics, the meaning of confusion matrix is needed. It is depicted in Fig. 6. Confusion matrix separates a dataset into four parts, positive samples predicted to be positive (TP), positive

Table 7 Experimental environment

System manufacturers	Dell
Processor	Inter Core i7-1065G7 CPU @1.30 GHz
Memory	8 GB
Operating system	Windows 10

Fig. 6 Confusion matrix

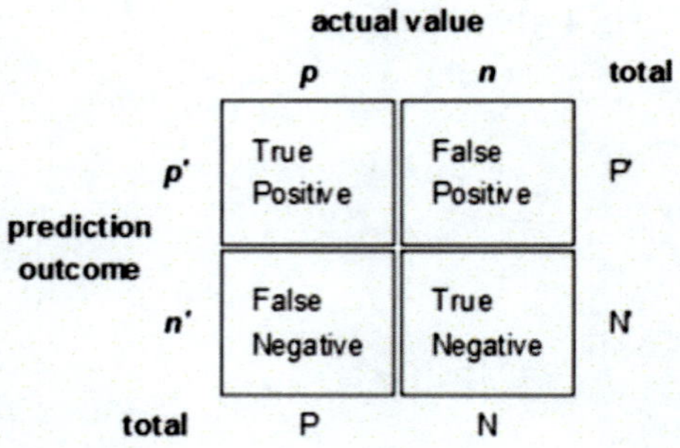

samples predicted to be negative (FN), negative samples predicted to be positive (FP) and negative samples predicted to be negative (TN).

AC means the proportion of truly-predicted samples in all samples. It can be calculated based on (2). PR represents the proportion of truly-predicted positive samples in all samples predicted to be positive, using (3). RC represents the proportion of truly-predicted positive samples in all positive samples, using (4). F1 score is derived from PR and RC, shown in (5).

$$AC = \frac{TP + TN}{TP + FN + FP + TN} \tag{2}$$

$$PR = \frac{TP}{TP + FP} \tag{3}$$

$$RC = \frac{TP}{TP + FN} \tag{4}$$

$$F1 = \frac{2 * PR * RC}{RC + PR} \tag{5}$$

In order to show the comparison of CNN model and machine learning measures, the train and test dataset are also utilized in SVM and Random Forest to train a specific model and get testing results. Selected metrics for evaluating CNN performance are also used in model above. In CNN-1 model and CNN-2 model, test dataset is divided into 10 epochs. Accuracy of every epochs are displayed in Figs. 7 and 8, which depicts the stability of CNN outputting positive results with high accuracy, above 98%. There is also a graphic view of accuracy gained from running different models on the test dataset shown in Fig. 9. CNN-1 and CNN-2 model impart an excellent role in DDos detection with 98.18 and 98.21% accuracy. Compared to them, Random Forest and SVM have slightly poorer performance.

Figure 10 demonstrates the comparison of detection precision of four models. CNN-1, CNN-2 and Random Forest have similar precision, which are close to 94% and better than SVM. Almost the same result is indicted in Recall and F1 score in Figs. 11 and 12.

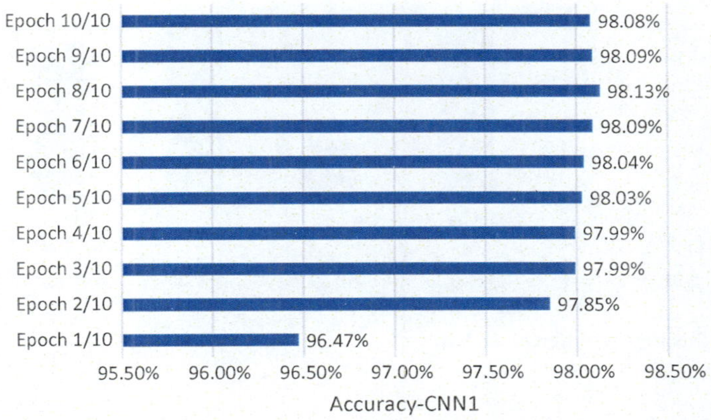

Fig. 7 Accuracy of different epoch in CNN1

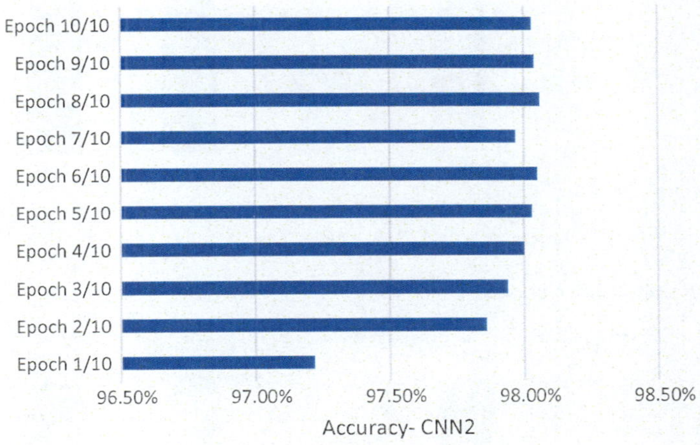

Fig. 8 Accuracy of different epoch in CNN2

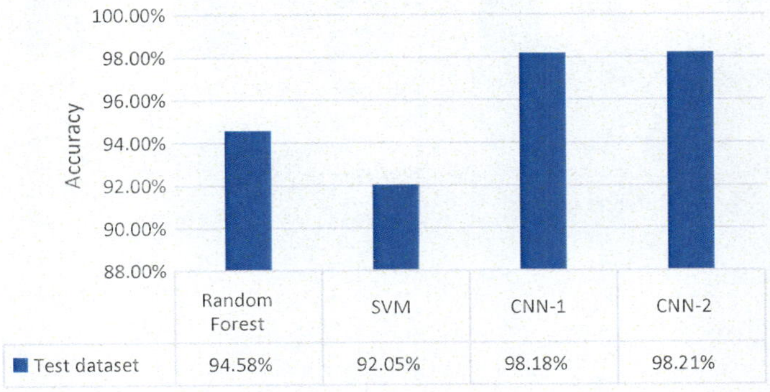

Fig. 9 Accuracy of different models

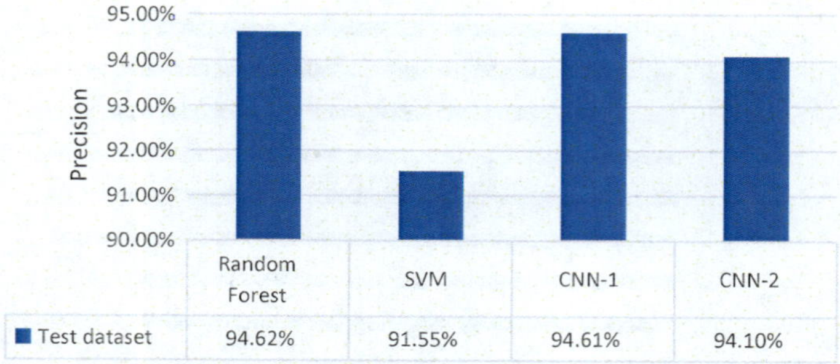

Fig. 10 Precision of different models

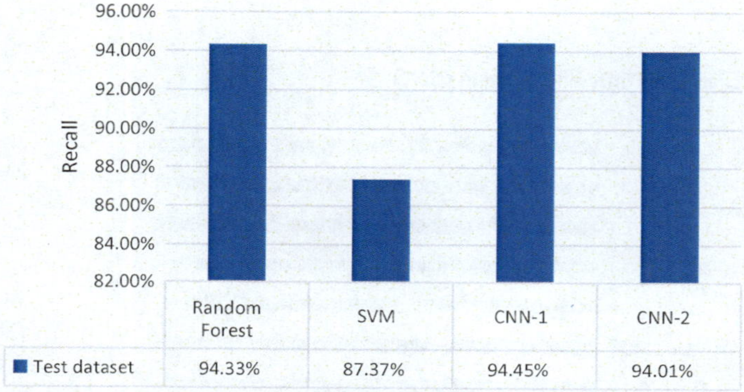

Fig. 11 Recall of different models

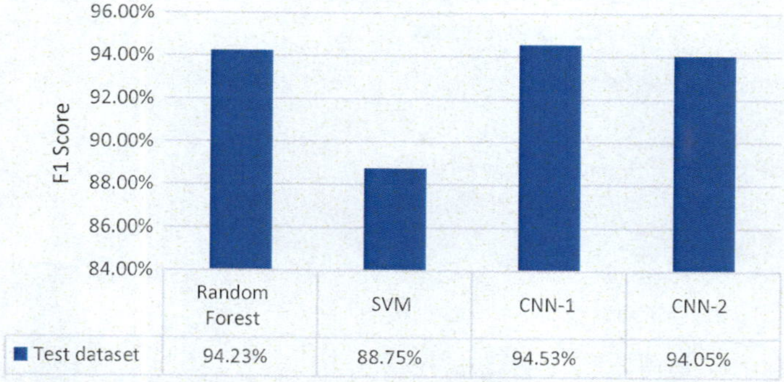

Fig. 12 Score of different models

6 Conclusion

Since the CNN model shows potential in intrusion detection, it is selected in this paper to conduct a DDos attacks classification. Five type of representative DDos attacks with modern reflective DDos attacks are included. CICDDoS2019 is the dataset analyzed in this paper. After data preprocessing, model constructing, training and testing, CNN demonstrates the best performance in accuracy compared to two other machine learning algorithms, SVM and Random Forest. For recall, precision and F1 score, CNN and Random Forest have similar numbers with high value, which is much better than SVM. CNN models with distinct multi-hidden layer amounts are applied to the datasets and show no difference in all metrics. CNN is endorsed as a suitable model for modern DDos attacks classification.

This paper makes a useful attempt. Our future work will focus on two aspects. One is to further optimize the model according to the characteristics of new dataset and attacks created in real network environment. The second is to further consider the time series of attack occurrence and improve the performance of the model in the actual environment of attack detection.

References

1. Zargar, S.T., Joshi, J., Tipper, D.: A survey of defense mechanisms against distributed denial of service (DDoS) flooding attacks. IEEE Commun. Surv. Tutor. **15**(4), 2046–2069 (2013)
2. https://www.cisco.com/c/en/us/solutions/collateral/executive-perspectives/annual-internet-report/white-paper-c11-741490.html. Accessed 20 Feb 2021
3. https://securelist.com/ddos-attacks-in-q1-2020/96837. Accessed 20 Feb 2021
4. https://aws-shield-tlr.s3.amazonaws.com/2020-Q1_AWS_Shield_TLR.pdf. Accessed 20 Feb 2021
5. Bace, R.G., Mell, P.: Intrusion detection systems (2001)
6. Jordan, M.I., Mitchell, T.M.: Machine learning: trends, perspectives, and prospects. Science **349**(6245), 255–260 (2015)
7. LeCun, Y., Bengio, Y., Hinton, G.: Deep learning. Nature **521**(7553), 436–444 (2015)
8. Wang, G., Hao, J., Ma, J., Huang, L.: A new approach to intrusion detection using artificial neural networks and fuzzy clustering. Expert Syst. Appl. **37**(9), 6225–6232 (2010)
9. Kokila, R.T., Selvi, S.T., Govindarajan, K.: DDoS detection and analysis in SDN-based environment using support vector machine classifier. In: 2014 Sixth International Conference on Advanced Computing (ICoAC), Dec 2014, pp. 205–210. IEEE
10. He, Z., Zhang, T., Lee, R.B.: Machine learning based DDoS attack detection from source side in cloud. In: 2017 IEEE 4th International Conference on Cyber Security and Cloud Computing (CSCloud), June 2017, pp. 114–120. IEEE
11. Kuang, F., Xu, W., Zhang, S.: A novel hybrid KPCA and SVM with GA model for intrusion detection. Appl. Soft Comput. **18**, 178–184 (2014)
12. Haider, S., Akhunzada, A., Mustafa, I., Patel, T.B., Fernandez, A., Choo, K.K.R., Iqbal, J.: A deep CNN ensemble framework for efficient DDOS attack detection in software defined networks. IEEE Access **8**, 53972–53983 (2020)
13. Chen, J., Yang, Y.T., Hu, K.K., Zheng, H.B., Wang, Z.: DAD-MCNN: DDoS attack detection via multi-channel CNN. In: Proceedings of the 2019 11th International Conference on Machine Learning and Computing, Feb 2019, pp. 484–488

14. Wu, K., Chen, Z., Li, W.: A novel intrusion detection model for a massive network using convolutional neural networks. IEEE Access **6**, 50850–50859 (2018)
15. Fu, Y., Lou, F., Meng, F., Tian, Z., Zhang, H., Jiang, F.: An intelligent network attack detection method based on RNN. In: 2018 IEEE Third International Conference on Data Science in Cyberspace (DSC), June 2018, pp. 483–489. IEEE
16. Albawi, S., Mohammed, T.A., Al-Zawi, S.: Understanding of a convolutional neural network. In: 2017 International Conference on Engineering and Technology (ICET), Aug 2017, pp. 1–6. IEEE
17. Sharafaldin, I., Lashkari, A.H., Hakak, S., Ghorbani, A.A.: Developing realistic distributed denial of service (DDoS) attack dataset and taxonomy. In: 2019 International Carnahan Conference on Security Technology (ICCST), Oct 2019, pp. 1–8. IEEE

KnowGraph-PM: A Knowledge Graph Based Pricing Model for Semiconductor Supply Chains

Nour Ramzy, Sören Auer, Javad Chamanara, and Hans Ehm

Abstract Semiconductor supply chains are described by significant demand fluctuation that increases as one moves up the supply chain, the so-called bullwhip effect. To counteract, semiconductor manufacturers aim to optimize capacity utilization, to deliver with shorter lead times and exploit this to generate revenue. Additionally, in a competitive market, firms seek to maintain customer relationships while applying revenue management strategies such as dynamic pricing. Price change potentially generates conflicts with customers. In this paper, we present KnowGraph-PM, a knowledge graph-based dynamic pricing model. The semantic model uses the potential of faster delivery and shorter lead times to define premium prices, thus entail increased profits based on the customer profile. The knowledge graph enables the integration of customer-related information, e.g., customer class and location to customer order data. The pricing algorithm is realized as a SPARQL query that relies on customer profile and order behavior to determine the corresponding price premium. We evaluate the approach by calculating the revenue generated after applying the pricing algorithm. Based on competency questions that translate to SPARQL queries, we validate the created knowledge graph. We demonstrate that semantic data integration enables customer-tailored revenue management.

N. Ramzy (✉)
Leibniz University of Hannover, Welfengarten 1, 30167 Hannover, Germany
e-mail: nour.ramzy@infineon.com

S. Auer · J. Chamanara
TIB Leibniz Information Centre for Science and Technology and L3S Research Center, Leibniz University of Hannover, Welfengarten 1B, 30167 Hannover, Germany
e-mail: auer@tib.eu

J. Chamanara
e-mail: chamanara@tib.eu

H. Ehm
Infineon Technologies AG, Am Campeon 1-15, 85579 Neubiberg, Germany
e-mail: hans.ehm@infineon.com

R. Lee (ed.), *Computer and Information Science 2021—Summer*,
Studies in Computational Intelligence 985,
https://doi.org/10.1007/978-3-030-79474-3_5

1 Introduction

The semiconductor industry is characterized by complex supply chain structures due to its wide range of customers with fluctuating demands for products, long production lead times, short product life cycles and rapidly changing technologies [12]. This entails specific phenomena such as the bullwhip effect that increases the variability of orders as they move up the supply chain from retailers to wholesalers to manufacturers to suppliers [17]. Consequently, matching the demand and capacity requires effective planning and optimal utilization of production capacities.

Moreover, manufacturing lead times are longer than customer order lead times (the time interval between order entry and requested delivery date), thus fulfilling order requests that go beyond promised delivery time is costly [14]. Therefore, to keep utilizing capacity in a profitable way and guarantee customer satisfaction, companies resort to revenue management ideas such as dynamic pricing [15], i.e., exploiting faster delivery to generate revenue. Potentially, conflicts with customers arise, affecting their satisfaction and harming the firm's long-term relationship with the customer and ultimately its success [20].

In fact, to maintain advantage especially needed within the competitive electronics market [11], semiconductor manufacturers strive to deliver and maintain various customer relationships, and design customized portfolios with distinct values to differentiate between customers' requirements [4]. Yet, revenue management and customer account management are often conducted independently of each other [19], although, the flow of information in a supply chain is crucial for carrying out effective exchanges between parties [16]. For instance, centralized information and elaborate information integration among supply chain partners can limit the increase in demand variability i.e., reducing the bullwhip effect [3].

In this article, we propose KnowGraph-PM a pricing solution to measure a price change to each customer for an order. The knowledge graph, depicting customers and orders, subsumes data about customer order behavior as well as customer profile. The former describes order lead times for a customer, e.g., requested, confirmed, while the latter defines customer classes expressing the relationship with the firm. For each customer, we calculate a price premium based on lead time variations as well as the corresponding price adjustment factor derived from the respective customer class. For each order placed by a customer, we apply the customer price premium to get the new price for this order. The approach enables, via a knowledge graph, semantic data integration by combining customer demand and order-related data with customer portfolio information. We apply the proposed approach to a use case of an international semiconductor manufacturing firm; the results show that KnowGraph-PM approach allows tailored revenue generation according to customers' profile thus reducing the risk of harming relationships.

The remainder of the paper is organized as follows: In Sect. 2 we introduce the domain and the key terminology and motivation behind the work. We present the approach and the KnowGraph-PM implementation in Sect. 3. In Sect. 4 we evaluate the implemented solution and discuss the advantages and limitations of our approach. Finally, we conclude and present an outlook for further steps in Sect. 5.

2 Background and Motivation

In this section we establish the context of the research. We explain the motivation for this research topic and introduce the terminology and essential to understanding the main aspects of the work.

2.1 Background

In this section we elaborate on some key terminology to understand the domain in question. This include concepts such as Lead Time and its variants, Lead Time-based Pricing, and Customer Relationship Management.

2.1.1 Lead Time

The *order lead time* (OLT) at each stage of the order, is the time frame between the entry of an order to the time point it passes a designated stage [13] as shown in Fig. 1.

A *requested order lead time* $OLT_{Requested}$ is the difference between *Customer Request Date* for an order, i.e., when the customer wishes to receive the order and the *Order Date* (when the customer placed the order). A *confirmed order lead time* $OLT_{Confirmed}$ is the difference between *Customer Delivery Date* i.e., the confirmed order delivery date and the *Order Date*. *SDT* is defined as the difference between *Standard Delivery Date* and *Order Date*. It is the lead time the customer can expect till the order is delivered. In practice, customers request their orders earlier than *SDT*, i.e., $SDT > OLT_{Requested}$. Moreover, due to the high-competitive market of semiconductors, manufacturers usually sell their products with a shorter $OLT_{Confirmed}$, closer to $OLT_{Requested}$. The potential of Lead Time-Based Pricing (LTBP) is seen

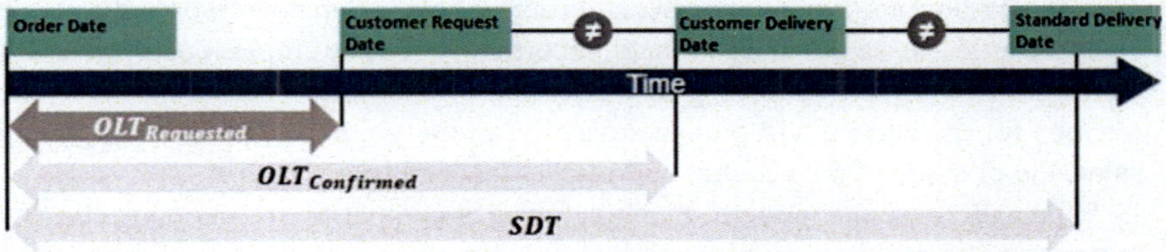

Fig. 1 The definitions of lead times

in the difference between *Standard Delivery Date* and *Customer Delivery Date*. Namely, faster delivery, $SDT > OLT_{Confirmed}$, is exploited by offering LTBP as part of dynamic pricing.

2.1.2 Lead Time-Based Pricing (LTBP)

Namely, dynamic pricing describes the firm's practice to charge various customers different prices for the same products [21]. For instance, [15] proposes a supply chain planning framework for revenue management that consists of solutions for demand steering and dynamic pricing where the pricing algorithm relies on the data from order lead time measurement. However, LTBP models to identify the price premium based on faster delivery are still under development. In fact, some rely on simple mathematical functions, e.g., linear, concave, convex. For instance, [2] introducing Eq. 1, explains the rationale behind a convex function; both opportunity costs for the manufacturer and for the customer increase with decreasing lead time.

$$P_{Premium} = \alpha * \log\left(\frac{OLT_{Confirmed}}{SDT}\right) \quad (1)$$

where α is a factor obtained from simulating customer order behavior for maximizing revenue for the company and $P_{Premium}$ is the price change or the added portion to the original price of the order. In fact, revenue management (RM) is built upon heterogeneity between customers; frequent buyers need to be treated differently than occasional ones [7]. Typically, Customer Relationship Management (CRM) enables customer profiling and delimits characteristics for customers and how they can be used to determine the price as part of RM [23].

2.1.3 Customer Relationship Management (CRM)

CRM refers to building one-to-one relationships with customers that can drive value for the firm [9]. Consequently, customization of requirements increases maintenance of customers [8]. Namely, by segmenting customers into portfolios, an organization can better understand the relative importance of each customer to the company's total profit [18]. Customer Portfolio Management (CPM) is based on financial performance as well as strategic and economic criteria, e.g., customer account types or classes, regional importance. For each criterion, the company chooses measurable characteristics to segment customers accordingly. For example, customer classes distinguish customers according to their impact on business, based on the average revenue for the current and previous fiscal year, the volume of purchases, potential sales, the prestige of the account and market leadership.

Especially in semiconductor manufacturing, a competitive domain, meeting customer specific requirements incites maintaining close associations with customers to identify their specific needs. In fact, CPM affects new product development [22].

Especially with the short product life cycle trait for this domain, it accentuates the need for manufacturers to act fast and maintain good relations with the customer.

2.2 *Motivation*

The bullwhip effect, a significant characteristic of the semiconductor supply chain, entails high inventory capacity, possibly unnecessarily. However, to keep utilizing capacity in a profitable way companies exploit revenue management ideas such as dynamic pricing. Yet, LTBP strategies are myopic because the decisions are optimized considering solely the profit that is expected to be obtained from the currently arriving customer without any foresight on the long-run impact of these decisions [13]. Companies realize the need to keep a good relationship with the customer while generating revenues.

Typically, interactions between revenue management and key account management have been largely ignored as both are often conducted independently of each other [19]. In addition, [10] explains how limited information sharing increases the difficulty of reducing the bullwhip effect and leads to inefficient supply chain management. Also, information integration increases acquisition and maintenance of customers according to profitability [8]. Moreover, to the best of our knowledge, previously developed LTBP models solely rely on customer order behavior while not including the customer portfolio and specific characteristics. In this work, we create a knowledge graph that semantically integrates data from different sources, i.e., customers, orders as well as customer account types. This solution suggests information integration from various data sources to customer-specific revenue management via LTBP. This improves revenue generation while maintaining customer relationships.

3 Implementation

This section describes the approach we implemented, integrating data from customer relationship management and customer order data.

3.1 *Knowledge Graph (KG)*

According to [6], a KG subsumes ontologies and instance data. In the following sections we introduce the domain ontology and the corresponding data sources.

3.1.1 Domain Ontology

We use the ontology-based data access (OBDA) approach for semantic data integration. The schema is given in terms of an ontology representing the formal and conceptual view of the domain [5]. Figure 2 shows the ontology comprising the *Order* class, which enables the coupling of the remaining classes of the domain, e.g., *Product* and *Customer*. The *Order* entity is uniquely identified via an *Order Number* and is described via twofold data properties. First, order information properties such as the order quantity and the original price of the order. Second, an order is described with lead time properties, e.g., order entry date, requested delivery date, and the standard delivery time. Through its two object properties *containsProduct* and *wasPlacedBy*, an order is linked to the *Product* and *Customer* entities respectively. The *Customer* class describes a customer, by assigning each customer a distinct customer code and by categorizing a customer into the defined customer account type or class. This categorization into a customer account type is then quantified with the data property *hasAdjustmentFactor*, which assigns each customer a specific pre-defined value based on their respective type. Finally, each product is uniquely identified with the help of the product number. Also, a *Product* includes data properties that contain all relevant product information such as the product basic type and the product line.

3.1.2 Data Sources: Use Case Description

The use case described here belongs to a international semiconductor manufacturing firm. We utilize these two datasets: *DS1* that contains data about order details, e.g., lead times, products and corresponding customers. We filter orders containing products within one product line. We process roughly 65 thousand orders in the time span of 2016-10-04 to 2020-09-01. *DS2* contains customer portfolio, i.e., customer account type, class and location. It subsumes 177 customers and corresponding account types (regular, key and others). Moreover, we use the corporate memory tool as a platform for data integration.[1]

3.2 Lead Time-Based Pricing Algorithm

The pricing algorithm is implemented as a SPARQL query that emulates the equation developed by [1]. SPARQL[2] is the query language standardized by the W3C for querying knowledge graphs. Figure 3 describes how parts of the equation match to the nested query shown. Steps [1–4] depict how the customer order behavior and customer account type contribute to calculating the price premium $P_{premium}$ for a

[1] https://eccenca.com/products/enterprise-knowledge-graph-platform-corporate-memory.
[2] https://www.w3.org/TR/rdf-sparql-query/.

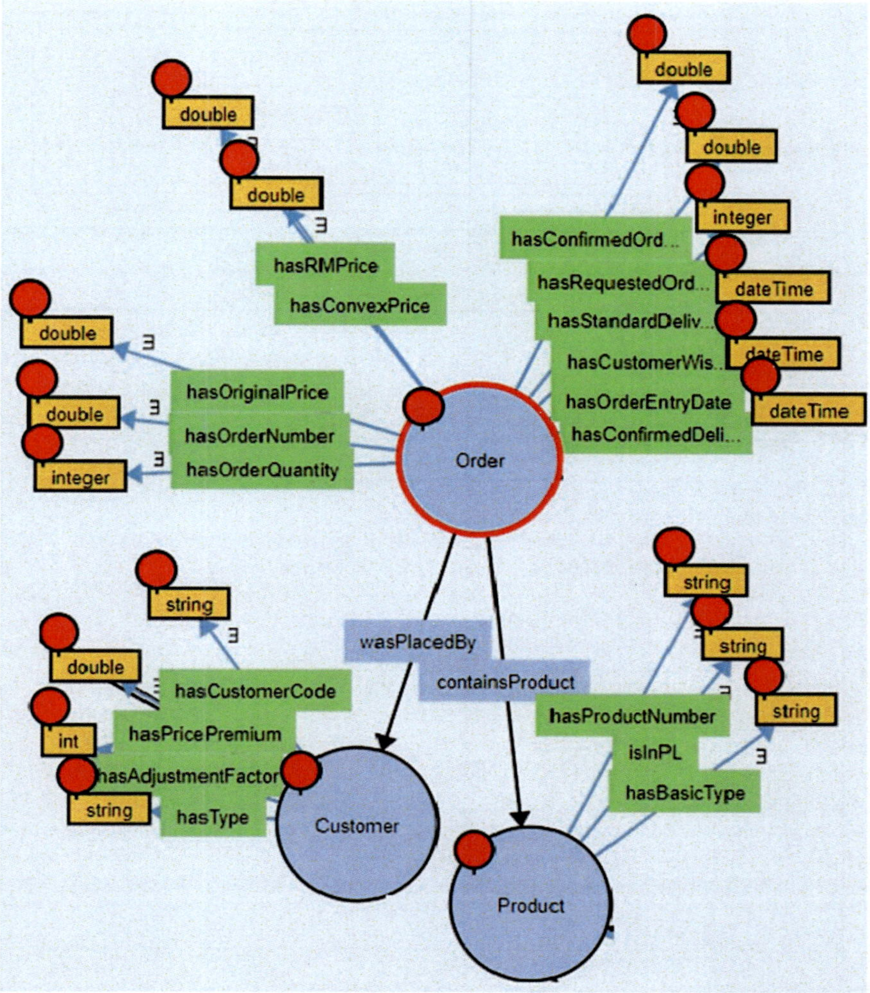

Fig. 2 Domain knowledge graph comprising the order, customer and product classes with associated properties

specific customer. The last step is to calculate the new revenue management price P_{RMj} of a specific order *j* knowing the $P_{premium}$ of this customer.

1. We calculate the relative standard deviation (RSD) and the relative mean deviation (RMD) for each customer individually. This is done only if revenue management is allowed, i.e., $SDT > OLT_{Requested}$, when customers request their orders earlier than *Standard Delivery Date*. This represents the customer order behavior from previous orders. α and β are coefficients to offer the model user the chance to emphasize one parameter more than the other. For instance, one could argue that

$$P_{premium_i} = Max\left\{\left\{Min\left[\left(\sqrt{\frac{\sum_{j=1}^{n}(x_j-\mu_i)^2}{n_i-1}}\right)*\frac{\alpha}{\mu_i}+\left(\frac{\frac{1}{n_i}\sum_{j=1}^{n}(x_j-\mu_i)}{|\mu_i|}\right)*\beta;(P_{max})\right]\right\}-\rho;1\right\}$$

RSD * α RMD * β ① ② ③ ④

```
INSERT { GRAPH <http://xyz.com/LTBP/> { ?customer :hasPricePremiun ?premium }}
WHERE                                                                          ④
  { GRAPH <http://xyz.com/LTBP/> {
  SELECT
    ?customer (IF (?price < 1, 1, ?price) as ?premium)                         ③
  WHERE
     {
    SELECT ?customer (IF (?rsd + ?rmd >= 2, 2 -?ro, ?rsd + ?rmd - ?ro) as ?price) ?rmd ?rsd ?ro
    WHERE                                                                      ②
      {
      SELECT(AVG(?mean) as ?m)(AVG(?r) as ?rc)
      ?customer (COUNT(?order) as ?totalorders)
      ((math:sqrt(SUM((?x - ?mean) * (?x - ?mean)) / (?totalorders - 1)) / ?m) as ?rsd)   ①
      (xsd:float(SUM(xsd:float(?x - ?mean)) / abs(?m)) as ?rmd)
      WHERE
        { ?order :wasPlacedBy ?customer.
          ?order :hasStandardDeliveryTime ?sdt.
          ?order :hasRequestedOrderLeadTime ?rolt.
          ?customer :hasAdjustmentFactor ?r.
          BIND(?sdt*7 - ?rolt as ?x). FILTER(?x > 0 ).
          {
        SELECT ?customer (AVG(?x) as ?mean)
        WHERE
          { ?order :wasPlacedBy ?customer.
            ?order :hasStandardDeliveryTime ?sdt.
            ?order :hasRequestedOrderLeadTime ?rolt.
            ?customer :hasAdjustmentFactor ?r.
            BIND(?sdt*7 - ?rolt as ?x) FILTER(?x > 0). }
        GROUP BY ?customer } }
        GROUP BY ?customer
        }
    }
  }}
```

Fig. 3 Lead time-based pricing algorithm with the formula (top) and corresponding SPARQL query implementation (bottom)

Table 1 Exemplary values of price adjustment factor

Revenue (r) range	Customer class	Price adjustment factor
$10 < r < 100$ m	Key	0.1
$5 < r < 10$ m	Regular	0.05
$0 < r < 5$ m	Others	0.025

the average deviation or mean is less important. For simplicity, alpha and beta are chosen to be equal to 1.

2. The sum of weighed RSD and RMD is then compared to a pre-determined threshold $P_{\max}$ which in this case is set to 2 as this represents the ceiling for the highest premium that can be offered to a customer which is double the original price. The sum is modified by the revenue adjustment parameter ρ. As shown in Table 1, for each customer account type or class there exists a corresponding ρ based on revenue ranges. Large revenue ranges, entail more important customer accounts, thus have greater advantage equivalent to higher ρ leading to smaller premium price.
3. In order to ensure that no customer receives a premium price smaller than the original we select the max between our first part of the formula and 1.
4. We assign $P_{premium}$ to each customer i where $1 < P_{premium} < 2$.
5. Upon a new order j, we apply Eq. 2 to get the revenue management price P_{RMj} from the original price P_{Oj} of the order and customer-specific $P_{premium}$ assigned in the previous step. As mentioned, the difference between SDT and $OLT_{Confirmed}$, is exploited for revenue management.

$$P_{RMj} = P_{Oj} * P_{Oj} \left[\left(1 - \frac{OLT_{Confirmed_j}}{SDTj} \right) * (P_{premium} - 1) \right] \qquad (2)$$

4 Evaluation

The implemented approach combines data from customer relationship management with customer order behavior. We use OBDA to represent the domain and create a knowledge graph. We evaluate KnowGraph-PM twofold. First, we check if the created knowledge graph covers the domain in question. Second, we calculate the total revenue generated using static and dynamic pricing algorithms and compare them with our KnowGraph-PM model.

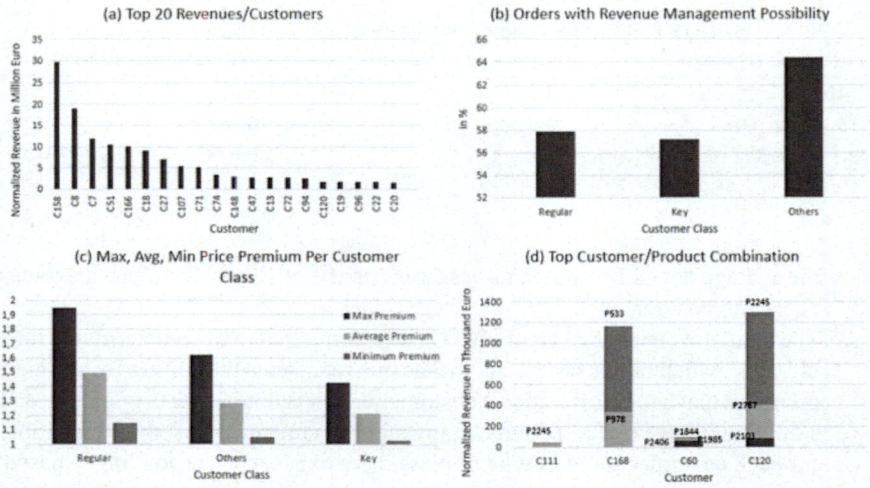

Fig. 4 Chart visualization of SPARQL query results

4.1 Competency Questions and KG Evaluation

We translate the listed competency questions, defined by domain experts, to SPARQL queries. We execute them on the described knowledge graph in Sect. 3. The results are extracted and represented in Fig. 4.

CQ1: Top 20 most profitable customers: This competency question, translated to SPARQL query in Listing 1, retrieves the top 20 customers that generate the highest revenues who are assigned a price premium and who buy the order with the price change. The results shown in Fig. 4a give a valuable insight into the most influential customers for a firm. This can be extended by examining these customer classes and accordingly tailor marketing strategies.

```
SELECT ?customer(SUM(?price) as ?TotalRevenue)
FROM <http://xyz.com/LTBP/>
WHERE {
    ?order :wasPlacedBy ?customer.
    ?order :containsProduct ?product.
    ?order :hasRMPrice ?price.
}
GROUP BY ?customer
ORDER BY DESC (?TotalRevenue)
LIMIT 20
```

Listing 1 SPARQL query of CQ1: top 20 most profitable customers

CQ2: Occurrence based customer ranking: We can examine the overall customer order behavior of the different customer classes. For this, we use the SPARQL query depicted in Listing 2 to rank customer types based on the highest percentage of occurrences of $SDT > OLT_{Requested}$ (RM is possible). We analyze, which customer classes are more likely to request a delivery date that is earlier than the *Standard Delivery Date*. This makes them eligible for revenue management with the adjustment factor ρ in Table 1 suitable to their respective customer class. The results of the query shown in Fig. 4b exhibit that *Key* account customers are least likely to request an early delivery. This is an expected outcome. In fact, key customers have strong relationship established with the company. Thus, they are more likely to enter long-term agreements and contracts entailing stable order behavior i.e., limited variation in lead times. Additionally, because of the frequency of their past orders, they can accurately forecast their needs and how a firm can fulfill them. Accordingly, this constrains dynamic pricing and revenue management potential.

```
SELECT ?type (COUNT(?temp) as ?RMPossible)
    (IF(?type = "Regular", ?RMPossible/r_count,
    IF(?type = "Key",
        ?RMPossible/k_count, ?RMPossible/o_count))
        as ?percentage)
FROM <http://xyz.com/LTBP/>
WHERE {
    ?order :wasPlacedBy ?customer.
    ?customer :hasType ?type.
    ?order :hasRequestedOrderLeadTime ?rolt .
    ?order :hasStandardDeliveryTime ?sdt.
    BIND((?sdt*7) - ?rolt as ?temp)
    FILTER(?temp > 0)
}
GROUP BY ?type
ORDER BY DESC (?percentage)
```

Listing 2 SPARQL query of CQ2: occurrence based customer ranking

CQ3: Per customer class price premium estimation: This question provides information about the customer class that is most likely to be the most profitable upon applying revenue management. This CQ is formulated as the SPARQL query in Listing 3. The information contains the highest, the lowest and the average Price Premium that will be paid by customers of each customer class. Also, it gives insights into the customer order behavior of the different customer classes. After executing the query, we observe in the results shown in Fig. 4c that the Regular customer type has the highest maximum, average, and minimum price premiums. This means that this customer class would yield the highest average revenue when the dynamic pricing model is applied.

```
SELECT ?type (MAX(?pprice) as ?MaxRM)
    (MIN(?pprice) as ?MinRM)
    (AVG(?pprice) as ?AverageRM)
FROM <http://xyz.com/LTBP/>
```

```
WHERE {
    ?customer :hasPricePremium ?pprice.
    ?customer :hasType ?type.
}
GROUP BY ?type
```

Listing 3 SPARQL query of CQ3: per customer class price premium estimation

CQ4: Initial customer and product selection: This CQ determines the combination of customers and products that should be released with the first practical implementation of the LTBP model in order to maximize the initial potential revenue increase using the adjusted prices. See Listing 4 for SPARQL implementation. The query results shown in Fig. 4d provide the combination of customers and products in terms of profitability and revenue generation in the case where the lead time-based pricing is applied.

```
SELECT  ?customer ?product
        (SUM(?price) as ?TotalProductRevenue)
FROM <http://xyz.com/LTBP/>
WHERE {
    ?order :wasPlacedBy ?customer.
    ?order :containsProduct ?product.
    ?order :hasRMPrice ?price.
    {
        SELECT ?customer ?pprice
        WHERE {
            ?customer :hasPricePremium ?pprice.
        }
        ORDER BY DESC (?pprice)
        LIMIT 5
    }
}
GROUP BY ?product ?customer
```

Listing 4 SPARQL query of CQ4: initial customer and product selection

4.2 *Revenue Management (RM)*

Dynamic pricing as part of RM is about generating revenue for the company. Existing dynamic pricing algorithms, e.g., convex model consider customer behavior while ignoring customer profiles. KnowGraph-PM customizes the RM price based on customer order behavior and customer relationship with the company. To evaluate this, we execute the query in Listing 5 where we sum the order prices with and without applying revenue management. The results show that $TotalOrginalPrice <$

$TotalRMPrice < TotalConvexPrice$. In that, *TotalOrginalPrice* is for the case that no RM is applied, which indeed is the summation of original prices for each order P_{Oj} as provided in Eq. 2, *TotalRMPrice* is the sum of RM price, i.e., P_{RMj} as of Eq. 2, and *TotalConvexPrice* is the sum of order prices if Eq. 1 is applied.

These results indicate that by applying KnowGraph-PM, a company can generate revenue while taking into consideration the customer portfolio. Other models such as the convex generate higher revenue but can entail a disturbed relationship with the customer.

```
SELECT
  (sum(?rmprice) as ?TotalRMPrice)
  (sum(?orignalprice) as ?TotalOrginalPrice)
  (sum(?convexprice) as ?TotalConvexPrice)
FROM <http://xyz.com/LTBP/>
WHERE {
  ?order  :hasRMPrice         ?rmprice.
  ?order  :hasOriginalPrice   ?orignalprice.
  ?order  :hasConvexPrice     ?convexprice.
}
```

Listing 5 Evaluation query

4.3 Discussion

The implemented approach integrates data from customer relationship management as well as customer order data to generate a price premium that fits the customer portfolio. The proposed solution is portable and can be applied to other domains with the right mapping between data and the ontology. Additionally, the two evaluation techniques show that it is up to the company to adapt pricing strategies based on customer portfolio. SPARQL queries can be tweaked to show details about a specific customer and consequently tailor marketing strategies.

However, the approach has some limitations. First, we identify a limitation in the evaluation as we compare our model to a convex model with α chosen to be -0.5 in Eq. 1. This choice is not accurate as the parameter is set after using simulation models to optimize the revenue for a specific set of customers. Second, for the evaluation, we calculate the total revenue with an assumption that all customers accept the price change. However, in practice, customers can refute the premium price and stick to the original.

5 Conclusion and Outlook

Semiconductor supply chains are strongly affected by the bullwhip effect, i.e., increasing demand fluctuation. Thus, in a competitive market, semiconductor manufacturers seek to optimize capacity utilization, deliver with shorter lead times than the agreed contractual times and guarantee customer satisfaction and loyalty. Firms exploit faster delivery by resorting to revenue management ideas such as dynamic pricing. However, this potentially affects relationships with customers.

In this work, we propose KnowGraph-PM, a knowledge graph lead time-based pricing approach allowing tailored revenue generation according to customers' profile thus reducing the risk of harming relationships. We present dynamic pricing as part of revenue management taking into consideration customer relationship with the customer. Using a pricing algorithm, implemented as a SPARQL query, we assign a price premium to each customer based on their order behavior and changes in the order lead time. Consequently, we calculate a price change for a customer's order. We apply the proposed approach to a use case of an international semiconductor manufacturing firm and proved that semantic data integration enables customer profiling aware lead time-based pricing.

As future work, we aim to integrate into KnowGraph-PM an acceptance rate distribution to model customer's behavior towards dynamic pricing. This would lead to more realistic figures in the evaluation thus overcoming the mentioned limitation. Moreover, we propose extending the knowledge-graph with more contextual data about the customer. For instance, Customer Master Data (address data, contract data) is an enabler for all customer-driven operations. Thus, it can be used to determine the strategic internal decision to treat the customer in a strategic manner, e.g., customer prioritization in times of allocation. Consequently, further pricing models that incorporate more factors about a customer can be explored.

Acknowledgements This work has received funding from the EU Electronic Component Systems for European Leadership (ECSEL) Joint Undertaking within Integrated Development 4.0 project (idev40) and the European Research Council for the project ScienceGRAPH (Grant agreement ID: 819536).

References

1. Berger, N.: Revenue management in the semiconductor industry (2017)
2. Berger, N., Strauss, A.: Order prioritization in the high-tech industry: a lead-time based pricing approach (2021)
3. Chen, F., Drezner, Z., Ryan, J.K., Simchi-Levi, D.: Quantifying the bullwhip effect in a simple supply chain: the impact of forecasting, lead times, and information. Manag. Sci. **46**(3), 436–443 (2000)
4. Chou, C.M.: How does manufacturing service perceived value influence customer satisfaction? An investigation of global semiconductor industry. Int. J. Prod. Res. **52**(17), 5041–5054 (2014)

5. De Giacomo, G., Lembo, D., Lenzerini, M., Poggi, A., Rosati, R.: Using ontologies for semantic data integration. In: A Comprehensive Guide Through the Italian Database Research Over the Last 25 Years, pp. 187–202. Springer (2018)
6. Galkin, M., Auer, S., Vidal, M.E., Scerri, S.: Enterprise knowledge graphs: a semantic approach for knowledge management in the next generation of enterprise information systems. ICEIS **2**, 88–98 (2017)
7. Hailwood, J., Gottlieb, D.: Customer relationship management. US Patent App. 09/972, 277 (2003)
8. Johnson, D.S., Clark, B.H., Barczak, G.: Customer relationship management processes: how faithful are business-to-business firms to customer profitability? Ind. Mark. Manag. **41**(7), 1094–1105 (2012)
9. Kumar, V.: Customer relationship management. In: Wiley International Encyclopedia of Marketing (2010)
10. Li, C.: Controlling the bullwhip effect in a supply chain system with constrained information flows. Appl. Math. Model. **37**(4), 1897–1909 (2013)
11. Mönch, L., Uzsoy, R., Fowler, J.W.: A survey of semiconductor supply chain models part I: semiconductor supply chains, strategic network design, and supply chain simulation. Int. J. Prod. Res. **56**(13), 4524–4545 (2018)
12. Mousavi, B.A., Azzouz, R., Heavey, C., Ehm, H.: Simulation-based analysis of the nervousness within semiconductors supply chain planning: insight from a case study. In: 2019 Winter Simulation Conference (WSC), pp. 2396–2407. IEEE (2019)
13. Öner-Közen, M., Ehm, H.: Dynamic price and lead time quotation under semiconductor industry related challenges. In: 2018 Winter Simulation Conference (WSC), pp. 3386–3396. IEEE (2018)
14. Ott, H., Heilmayer, S., Sng, C., et al.: Granularity dependency of forecast accuracy in semiconductor industry. Res. Logist. Prod. **3** (2013)
15. Seitz, A., Ehm, H., Akkerman, R., Osman, S.: A robust supply chain planning framework for revenue management in the semiconductor industry. J. Revenue Pricing Manag. **15**(6), 523–533 (2016)
16. Stefansson, G.: Business-to-business data sharing: a source for integration of supply chains. Int. J. Prod. Econ. **75**(1–2), 135–146 (2002)
17. Sucky, E.: The bullwhip effect in supply chains—an overestimated problem? Int. J. Prod. Econ. **118**(1), 311–322 (2009)
18. Thakur, R., Workman, L.: Customer portfolio management (CPM) for improved customer relationship management (CRM): are your customers platinum, gold, silver, or bronze? J. Bus. Res. **69**(10), 4095–4102 (2016)
19. Wang, X.L., Brennan, R.: A framework for key account management and revenue management integration. Ind. Mark. Manag. **43**(7), 1172–1181 (2014)
20. Wirtz, J., Kimes, S.E., Theng, J.H.P., Patterson, P.: Revenue management: resolving potential customer conflicts. J. Revenue Pricing Manag. **2**(3), 216–226 (2003)
21. Wittman, M.D., Belobaba, P.P.: Customized dynamic pricing of airline fare products. J. Revenue Pricing Manag. **17**(2), 78–90 (2018)
22. Yli-Renko, H., Janakiraman, R.: How customer portfolio affects new product development in technology-based entrepreneurial firms. J. Mark. **72**(5), 131–148 (2008)
23. Zhang, M., Bell, P.: Price fencing in the practice of revenue management: an overview and taxonomy. J. Revenue Pricing Manag. **11**(2), 146–159 (2012)

A Study on the Recognition of Hangeul Through Transitional Learning in Handwritten Application

Jae Hyuk Heo, Sang Wook Lee, Hee Won Lee, and Gwang Yong Gim

Abstract Through the utilization and convergence of technologies, such as artificial intelligence, big data, and the Internet of Things, that can represent the fourth industrial revolution, today's world is changing rapidly. However, public organizations in Korea are still manually entering reports written by hand in many fields into the information system by a government official in charge. This study studies the improvement method for optical character recognition (OCR) of hand-written General Educational Development (GED) application. Through learning using cursive fonts, separating images for items subject to information service, and managing whitelist by item, the recognition rate is to be improved. In the field of optical character recognition (OCR), the recognition rate of printed texts written in Hangeul font is over 90%, but the recognition rate of cursive writing is very low. Various tables and letters of different sizes in hand-written application form documents hinder the recognition rate. Although the recognition rate has improved a lot recently with various studies using Deep Learning, there is a problem that takes too much processing time. Of the 11,172 Hangeul characters, 2350 are used. On the other hand, English has 52 case-sensitive characters. The recognition rate of Hangeul, which consists of a combination of initial sound, medial vowels, and final consonant, is bound to drop significantly. As mentioned earlier about a performance issue, software with open source-based neural network learning engines was used. In addition, the status of Hangeul used in names and addresses was investigated for the application of whitelists. The cursive font learning, image separation and whitelist proposed in this study have been effective in improving recognition rates.

Keywords Hangeul handwriting · White-list · OCR · Artificial neural network

---

J. H. Heo · S. W. Lee · H. W. Lee
Department of IT Policy and Management, Graduate School, Soongsil University, Seoul, Korea
e-mail: picro@daum.net

G. Y. Gim (✉)
Department of Business Administration, Soongsil University, Seoul, Korea
e-mail: gygim@ssu.ac.kr

R. Lee (ed.), *Computer and Information Science 2021—Summer*,
Studies in Computational Intelligence 985,
https://doi.org/10.1007/978-3-030-79474-3_6

1 Introduction

Through the convergence of technologies that can represent 4th Industrial Revolution such as artificial intelligence, big data, and the Internet of Things, the whole world is changing rapidly. Korea is also changing the way it works fast due to the introduction of 52 h a week and a decrease in the number of people who are able to produce and experiencing an increase in the number of senior citizens. Not only that, there has been a change in perception that values work-life balance. These effects have increased the importance of digital business processing [1].

Korean public organizations are still receiving documents hand-written by civil petitioners for various reports and applications. These documents include the Korea Customs Service's import/export declaration and customs declaration, the Ministry of Justice's (entry and exit management office), the Ministry of Land, Infrastructure and Transport's real estate transaction report, the Ministry of Public Administration and Security's birth report, and the city and provincial education offices' application for General Educational Development (GED). A public official in charge is manually entering the application into a public information system operated by each ministry.

Applications for the GED application by the city and provincial education offices, which are subject to this study, are received online. However, about one-third of the applications have been written and received manually, and the officers in charge enters the application information into the information system.

This study presents the Korean Optical Character Recognition (OCR) method after scanning in the form of an electronic file for the information of the written application for the GED.

A lot of research is also being conducted on the Korean Optical Character Recognition (OCR) using Deep Learning technology. As a result of the related research and analysis, the recognition rate of Hangeul has improved a lot, but the processing time has been excessive, so there is a limit to applying it to the actual work.

In this study, the open-source software Tesseract [2, 3], which shows rapid processing speed, was used to conduct research on Korean optical character recognition (OCR). Tesseract 4 has added a new neural network (LSTM) based OCR engine that focuses online recognition [4].

2 Research Background

2.1 Overview of Optical Character Recognition (OCR)

Optical character recognition (OCR) is the conversion of human hand-written or printed documents into electromagnetic waves through an image scanner to a computer-recognized character code.

OCR began in 1928 when German G. Taushek patented methods for character recognition using pattern matching techniques. Currently, much attention is being paid to research using artificial neural network models. It is used for recognizing vehicle numbers of parking lots and speeding cameras and recognizing text messages of pictures or images on smartphones.

The recent use of optical character recognition (OCR) can be divided into four ways. Artificial neural network methods, structural analysis methods, statistical methods, and template matching [1, 5]. Recently, research using the Neural Networks Model in artificial intelligence systems for recognition of character patterns has received the most attention and is being used.

2.2 Whitelist

The meaning of the whitelist is to allow only what has been proven to be 'safety', in contrast to the blacklist that blocks "maliciousness" proven. This concept is mainly used in security. Instead of whitelists and blacklists, they also speak as a way of security called 'positive' and 'negative' [6]. It is expected that this whitelist method will be applied to optical character recognition (OCR) to improve recognition rate. If the original document operates a whitelist for each item needed for information, the number of Korean or English words used for each item will be limited. In this study, the whitelist design is embedded to be applied to names and addresses.

2.3 Recognition of Korean Letter

Research on Korean optical character recognition (OCR) is being conducted in various ways. The research by the Korean Optical Character Recognition (OCR) focuses on the study of the separation of consonants and vowels and the structural characteristics of Hangeul itself, thus separating them into letter units. As deep learning technology has become more common, research on recognition of letter unit is being actively carried out rather than the recognition of separation of consonants and vowels. The characteristics of the documents written in Korean are the combination of various types of characters such as Hangeul, English characters, Chinese characters, numbers, and symbols. In the case of English, it consists of 52 independent alphabets, from lowercase letters a to z uppercase A to Z and is used independently to form words. On the other hand, Hangeul consists of a combination of 19 initial consonants, 21 medial vowels, and 28 final consonants.

In Hangeul, the total number of possible combinations is 11,172 characters, and the actual use of Hangeul is known to be 2350 characters. Hangeul will inevitably have a lower probability of recognition than English recognition in a letter-based recognition system. In addition, the combination of consonants, mediality, and longitudinal increases the degree of similarity between letters, which increases the complexity for recognition. Character recognition through deep learning is a model that can find optimal distinctions by deriving various characteristics of Hangeul [7]. However,

Table 1 Comparison of Tesseract performance time with CNN [11]

Dataset	Method	Process time (s)
LG	CNNs	14,032
T.I.G	Tesseract	26
N.I.G	CNNs	14,183
	Tesseract	165

when there are many similarities between letters and the number of combinations, the generalized performance for new data is significantly reduced [7].

Kim [8] raised recognition rate of the pre clustered Hangeul image by using the CNN model. In addition, Kim and Cha [9] improved the recognition performance of Hangeul images by 89% through a simplified GoogLeNet study for Hangeul recognition. Ko [10] proposed a faster and more accurate model of Hangeul recognition than before by conducting experiments with 230,000 Hangeul font images through simplified VGGNet [8–10]. Kim and Cha et al. [9] obtained more accurate data and conducted CNN model learning through the thinning algorithm for preprocessing [9]. The high recognition rate is focused on research through deep learning. Ko [11] studied the recognition rate and performance of Deep Learning's CNNs and open-source software TesserAct [3, 11, 12].

In Table 1, it took 14,183 s for CNNs to recognize N.I.G Dataset, and 165 s for Tesser-act, about 90 times the difference. Figure 1 showed a whopping 53% difference between CNNs (94%) and Tesseract (41%) [11].

N.I.G Datasets are 5295 characters that randomly insert impulse or Gaussian noise for the characters 0 to 9, A to Z, a to z, [ ] and ( ) [11].

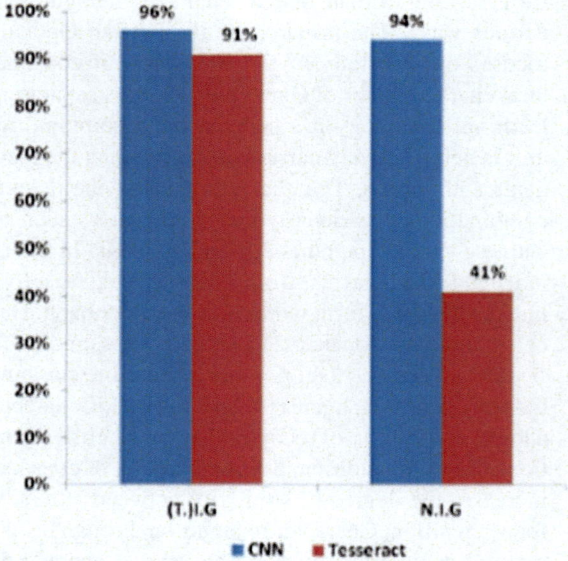

Fig. 1 Comparison of recognition rate between CNN and Tesseract [11]

In this study, we trained Tesseract using Hangeul cursive font for the efficiency of Hangeul recognition. In addition, target items that require informationization are separated into separate images to proceed with recognition. In addition, it was intended to confirm the impact on the recognition rate by providing a whitelisted list of the characters used in the item.

2.4 Korean Name

It is common for a Korean name to be composed of three syllables, including first and last name. However, in some cases, it may be prolonged due to the intention of making a unique name.

Family register established rule No. 651 <Guidelines for family register office regard to record letter of names> 11 stipulates that the names of Koreans shall be not more than six characters in total, including their last names [13].

In the case of Korean names, male names made in two letters are 98.33%, and female names are 98.84%, accounting for a high proportion of more than 98% of the total. Except for the special case of Korean names, most of them consist of two syllables [13]. Korean names are changing their preferred characters according to the times, but they prefer certain characters.

The statistics on names are ranked from 2008 to 2019 among the top 1500 male and female names through the statistical service of the electronic family relationship system of the Supreme Court. In the electronic family relation system, 118 surnames accounted for 99.9% of the total family names from 2008 to 2019. The total number of Korean letters used in 118 last names and the top 1500 names for men and women is 298. It is estimated that only 12.7% of the 2350 Korean characters used in the name are used.

2.5 Korean Address

The Korean address consists of several levels of hierarchy according to the administrative district system. There are provinces, special cities, and metropolitan cities in the top tier, and counties, cities, and districts in the next tier. There is a dong under the Gu and city, and Eup, and Myeon are under the county.

Large cities, such as Bundang-gu in Seongnam City and Manan-gu in Anyang City, also have a district under them. Next, it consists of Dong, Eup, Myeon, Li, and Dong [14].

Reference was made to the type and frequency survey of the characters used in the first and second half of the address. A general table of letters used in the first and second half of the address [14]. A survey of the type and frequency of characters used in addresses confirmed that only 403 Korean characters are used [14].

Although the road name address has been in effect since 2014, the road name address also uses the first half of the existing address and the latter part is used in the form of OO-gil after the existing name. Therefore, it is expected that the content surveyed will not deviate significantly.

2.6 Transfer Learning

What is important in neural network learning is enough data to distinguish characteristics for various learning targets. When there is enough good learning data to characterize the target, neural networks are more likely to show good performance through learning. However, collecting a lot of quality learning data is difficult in reality.

Transfer learning is the way to solve this lack of learning data. By using transition learning, even when solving already learned problems and other problems, some or all the neural networks learned with rich and good data can be reused instead of building models from scratch. The neural network already has well-learned parameters, so the learning starting point is good.

Computer vision systems using transition learning can achieve high accuracy in a relatively short time, reducing the development time of the plan [15].

3 Research Method

In this study, a measure is proposed for efficient Hangeul letter recognition for written GED applications. There are three main factors for improving handwriting recognition.

First, separate the image for the object to be digitized in GED application. Separate images are designated as optical character recognition (OCR) languages (Korean letters, numbers) according to the information target (name, resident registration number, residence, telephone number, mobile phone).

Second, it manages whitelists by information target. Specify resident registration numbers, phone numbers, numbers and characters that can appear on mobile phones, and designate a list of Korean characters that can appear in names and residences.

Third, use learned font data as transfer learning data to proceed with learning additional cursive font.

Through the three improvement measures, we want to improve the recognition rate of optical character recognition (OCR).

Fig. 2 Tesseract processing procedure

3.1 LSTM-Based Open-Source Software Tesseract

Tesseract [3, 12] is an open source optical character recognition (OCR) engine based on command line interface. Because API is provided, it can be used in languages such as C++, Python, and JAVA. In the beginning, it was recognized only for English, but through machine learning, it supported more than 60 languages and provided learning methods to recognize new languages.

In the beginning, it recognized only for English, but through machine learning, it has been supporting more than 60 languages and providing learning methods to recognize new languages.

Long Short-Term Memory (LSTM) based engines are added and Unicode is supported to recognize for a total of 116 languages since 2015.

It provides learning data files for each language and is an archive file in its own format [4].

For Latin-based languages, it provides learned data for about 400,000 text lines for 4500 fonts [4]. The advantage of Tesseract is to provide LSTM-based learning capabilities for different fonts in different languages. Character recognition has relatively fast performance, but learning can take days or weeks.

Regarding the training method, there is a method of training from the beginning regarding that language or new training from top layers or any layer. Also, there is a fine-tuning method of learning or re-learning similar existing training data, but with new unusual fonts.

The operation of the Tesseract is as shown in Fig. 2. The image subject to recognition is treated as input to the optical character recognition (OCR) engine. After binarizing the input image using the threshold values, the components of the binary image are analyzed, and the outline is extracted. After analyzing and organizing text lines, separate them into words according to the letter's spacing. Recognize content divided by words and pages.

3.2 Hangeul Training Data

In order to recognize cursive Korean, best (most accurate) LSTM models (kor.traineddata) made by training Hangeul fonts from Tesseract were used as transfer learning data to conduct learning on cursive fonts. The learned cursive fonts used 109 cursive fonts distributed by Naver [16]. The distributed handwriting font embodies the handwriting characteristics of various people.

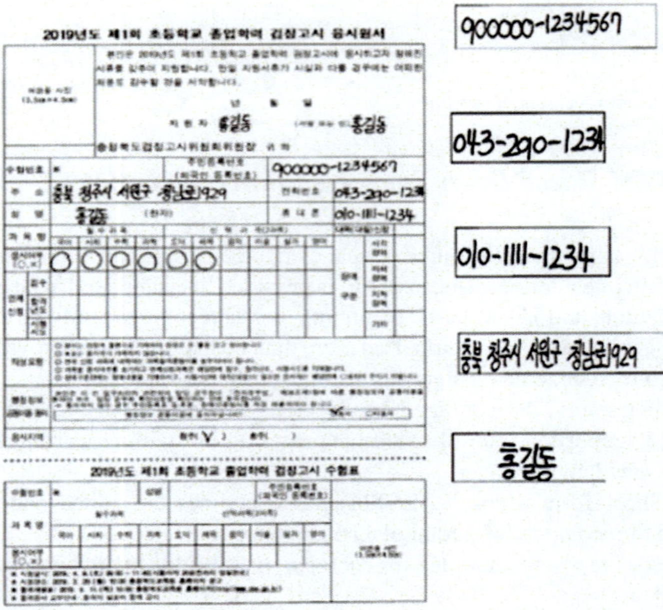

Fig. 3 GED application

3.3 Separation of the Data of the GED Application Form

The application for the qualification exam is as shown on the left side of Fig. 3.

The right is an image of an itemized name (applicant), resident registration number (foreign registration number), address, phone number, and mobile phone.

3.4 Whitelist Application

The processing process for whitelists is as shown in Fig. 4. It indicates the processing process of optical character recognition (OCR) of image '1'. Assuming a 60% probability of '/' being '/' for image '1', 30% probability of '1' and 10% probability of 'l'

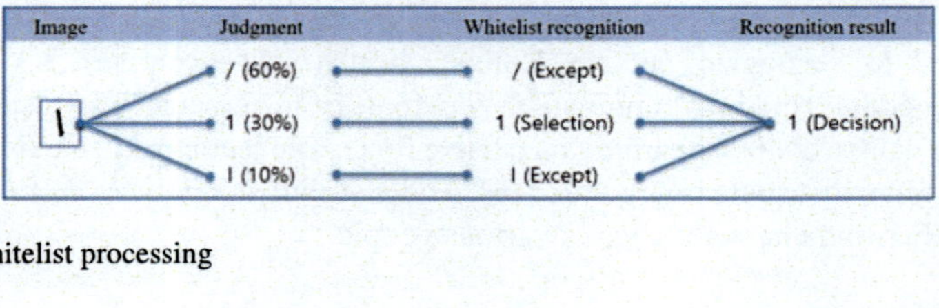

Fig. 4 Whitelist processing

for image '1', if the whitelist was specified as '1234567890-', the final recognition result would be '1' contained in the whitelist rather than '/' with the highest probability of '/'. The whitelist is specified for each item. The number of resident registration numbers, phone numbers and mobile phones will be designated as whitelists, with 403 Korean characters used in their addresses and 298 Korean characters used in their first and last names.

4 Experiment Result

This experiment was conducted in a KVM (Kernel-based Virtual Machine) environment.

During optical character recognition (OCR), the recognition of the entire document was completed within 4 s and the image of each item was all completed within 1 s, so no performance issue occurred.

The hardware is KVM Processor (2.0 GHz, 2 Core), 8 GB DIMM RAM, and HDD 1 TB. OS is Ub-untu 20.04 OCR is tess4j 3.4.8 (Tesseract 4.1 for Java), JDK 1.8.

The learning data used kor of the best (most accurated LSTM models) provided by GitHub Testeract.

Based on existing learning data, 400,000 repeated lessons were conducted on 109 new shared cursive fonts, resulting in error rates of 0.216%/0.473%/1.346% (Mean RMS/CER/WER) (Table 2).

The contents of resident registration numbers, addresses, names, phone numbers and mobile phones written in the application form for the GED correspond to personal information, but personal information was not entered in this experiment.

The same content (name: Hong Gil-dong, resident registration number: 900000-1234567, address: 1929 Cheongnam-ro, Seowon-gu, Cheongju-si, Chungbuk, telephone number: 043-290-1234, mobile phone: 010-1111-1234) was written differently and the recognition was tested.

100 people were required to fill out the same application form for the qualification exam.

The experimental sequence was carried out in the same order as the one shown in Fig. 5.

Table 2 Training result

Iteration	Mean RMS (%)	Delta (%)	CER (char error rate) (%)	WER (word error rate) (%)
100	1.648	7.614	18.875	46.192
1000	1.322	5.019	13.271	33.575
10,000	0.759	1.805	4.969	13.571
200,000	0.296	0.325	0.911	2.610
400,000	0.216	0.169	0.473	1.346

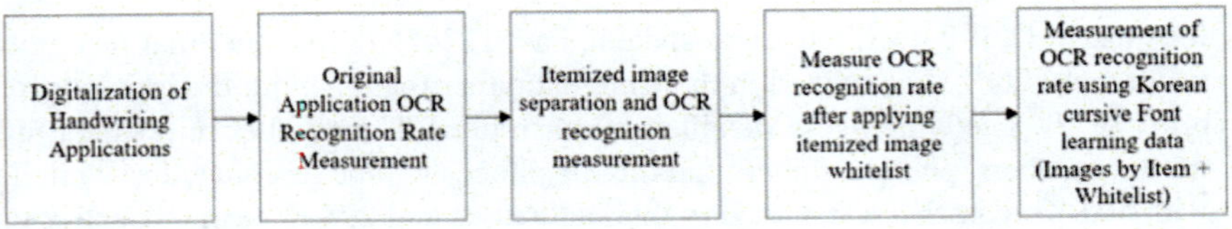

Fig. 5 Experimental sequence

The Korean language recognition was conducted for 100 written test applications. Figure 6 is an example of the progress of the Korean language recognition process on the application for the GED.

Figure 6 shows Original Application OCR, Itemized image division and OCR, OCR after applying itemized image whitelist and OCR using Korean language data (Imagine Whistling + Item).

OCR (itemized + Whitelist) using handwriting font learning data shows the best recognition rate.

Figure 7 is a Korean-language recognition graph for 100 original documents. RR (Recognition Rate). (A) is the OCR of the Original Application. RR. (B) is the OCR

Item	Original Data	Original OCR	Itemized OCR	Whitelist OCR	OCR using Learning Data (itemized + Whitelist)
Name	홍 길동	[illegible]	[illegible]	[illegible]	홍 길 동
Resident registration number	[illegible]	900000 - 123 567	900006 - 123 5677	900006-123 5677	900000-134 567
Residence	[illegible]	[illegible]	[illegible]	[illegible]	[illegible]
Telephone number	043 - 290 - 1234	2042 - 2926 -/224	0463 - 29?6 - /224	0463-2976-224	043-290-1234
Cell phone number	010 - 1111 - 1234	12/8-////-/236	2/2 - ////-/236	22--7236	010-1111-1234

Fig. 6 Example of experimental results

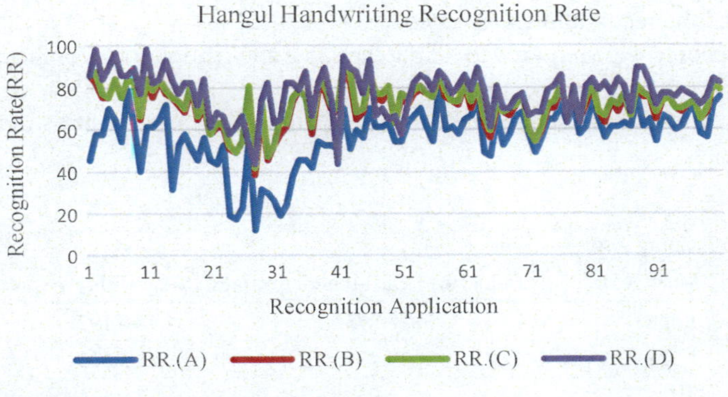

Fig. 7 Results of processing after whitelisting

of the Itemized image division. RR. (C) is an OCR after applying itemized image whitelist. RR. (D) is an OCR using Korean cursive Font leading data.

The recognition rate using the Korean cursive Font warning data has improved. The overall average recognition rate improved to 56.3% for RR (A), 71.4% for RR (B), 72.8 for RR (C) and 77.6% for RR (D).

5 Conclusion

5.1 Research Summary and Implication

In this study, a study model was presented for recognizing cursive in hand-written test applications. With the advent of the fourth industrial revolution technology era, such as artificial intelligence, big data, and the Internet of Things, the scope of interest and utilization of optical character recognition (OCR) is expanding. Various projects are under way, and research on the recognition of Hangeul in various ways is being carried out continuously. Languages that divide consonants and vowels into alphabetical sequencing, such as English, by many studies and data scale, have more than 97% recognition rates in print and cursive writing. However, characters made up of combinations of initial consonants, medial vowels, and final consonant, such as Hangeul, still lack recognition rates. In addition, the recognition rate of Korean handwriting using deep learning is good, but it takes a long time to process, making it difficult to use it practically. In this study, a cursive font recognition model was presented for recognizing the hand-written qualification exam application using Tesseract which LSTM-based open source software. In order to verify the improvement of the proposed model for the GED application written in 100 handwriting, whitelist application of images separated from the entire optical character recognition (OCR) processing in the original document, and the recognition rate was measured through cursive font learning data.

Experiment result have shown that the model proposed in this study can improve the targeted cursive recognition rate. The recognition rate due to item separation was improved by 15.1%, and the recognition rate was improved by 1.5% more than item separation by whitelist application. The recognition rate improved by 4.8% more after learning Korean cursive font.

Tesseract tests the text lines to determine whether they are fixed pitch. Where it finds fixed pitch text, Tesseract chops the words into characters using the pitch, and disables the chopper and associator on these words for the word recognition step. The left side of Fig. 8 shows a typical example of a fixed-pitch word [2].

Hangeul is Pig. It is difficult to judge the beginning and end of the correct character by inter-character interference, as on the right side of 10. And because of the confusion in the perception of 'ㅇ' and 'ㅁ', 'ㄷ' and 'ㄹ' due to spilling, it is difficult to find fixed-pitch, which also affects the perception of Hangeul.

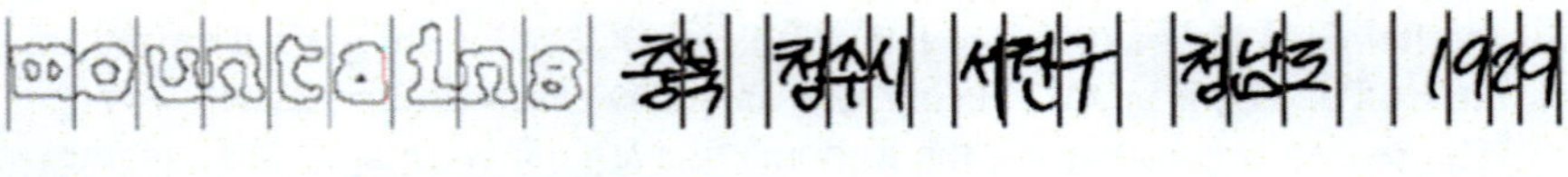

Fig. 8 A fixed pitch chopped word by Smith [2]

5.2 *Future Challenge and Discussion Topics*

Latin-based learning data were learned on more than 3000 fonts. The recognition rate of cursive writing in English averages 97%. Handwriting recognition through deep learning, which has been studied a lot recently, has a disadvantage of high recognition rates but excessive processing time [11].

On the other hand, Tesseract's Korean handwriting recognition process is fast, but its recognition rate remains at 77.6%, making it difficult to actually use it. Further research is needed to ensure the processing speed and recognition rate of Korean cursive optical character recognition (OCR) (more than 90%).

References

1. Sung-Yong, Y.: Automated design of unstructured text image data based on machine learning powered robotic process automation. Doctoral dissertation, Hoseo University, Seoul, Korea (2018)
2. Smith, R.: An overview of the Tesseract OCR engine. In: Ninth International Conference on Document Analysis and Recognition (ICDAR 2007), vol. 2, Sept 2007, pp. 629–633. IEEE
3. Patel, C., Patel, A., Patel, D.: Optical character recognition by open source OCR tool Tesseract: a case study. Int. J. Comput. Appl. **55**(10), 50–56 (2012)
4. Tesseract-ocr tessdata-best [website], 5 Oct 2020. https://github.com/tesseract-ocr/tessdata_best
5. Cho, M.H.: A study on character recognition using wavelet transformation and moment. J. Korea Soc. Comput. Inf. **15**(10), 49–57 (2010)
6. GooChoon, J.: The architecture design of automated whitelist verification system for secure mobile E-gov service. Master dissertation, Soongsil University, Seoul, Korea (2014)
7. SangUk, H.: Detection and extraction of objects in a document using object detection method. Master dissertation, Kookmin University, Seoul, Korea (2019)
8. Kim, J.Y.: Hangeul handwriting recognition using multi-columned deep learning structure and self organizing map. Master dissertation, Soongsil University, Seoul, Korea (2018)
9. Kim, Y.G., Cha, E.Y.: Streamlined GoogLeNet algorithm based on CNN for Korean character recognition. J. Korea Inst. Inf. Commun. Eng. **20**(9), 1657–1665 (2016)
10. Ko, W.: Recognition and classification of Korean fonts based on CNN for digital forensics. Master dissertation, Ewha Womans University, Seoul, Korea (2019)
11. Ko, D.: Optical character recognition performance comparison of CNNs and Tesseract. Maser dissertation, Sungkyunkwan University, Seoul, Korea (2016)
12. Slides from Tesseract tutorial at DAS, Santorini [website], 12 Sept 2020. https://github.com/tesseract-ocr/docs/blob/master/das_tutorial2016
13. Hui-suk, G., Myung-hee, Y., Dong-geun, P.: A sociolinguistic study on characteristics of names of Korean people and their transitional aspects. Lang. Study **73**(73), 33–60 (2012)
14. Kim, M.K., Kwon, Y.B.: A frequency measure of hangeul in Korean zip code. In: Annual Conference on Human and Language Technology, pp. 295–301 (1993)

15. Rawat, W., Wang, Z.: Deep convolutional neural networks for image classification: a comprehensive review. Neural Comput. **29**(9), 2352–2449 (2017)
16. New Nanum handwriting font-Naver Clova [website], 4 Sept 2020. https://clova.ai/handwriting

Study on Partial Image Detection for Drawing—Focus on Unstructured Images Included in the Main Image

Ji Won Lee, Jae Ho Lee, Doh Yeon Kim, and Gwang Yong Gim

Abstract AI is being used in various fields, especially in image analysis. There are many ways to analyze images using AI. There are many cases of image analysis that recognize real objects, but it is difficult to find application cases for image analysis cases that are not common. Unusual objects, that is, drawing images, such as hand-drawn drawings, are rarely learned and have limitations in image detection. In order to overcome these limitations, this study aims to learn images by applying deep learning so that they can be applied to picture projection tests among unusual fields, and to detect characteristic images that can be used for psycho-logical interpretation. In order to detect atypical images that may have psycho-logical characteristics among pictorial images, we learned through image trans-formation and increase. As a result of testing whether it is possible to detect non-realistic images through deep learning, first, it is possible to detect images even in hand drawings that are not real photos, and second, it is possible to detect unpatented partial images in pictures that must be found important for psychological analysis. It was confirmed that. Predicting the state of mind due to the detection of characteristic partial images Based on this paper, it is expected that faster and higher-level psychological analysis will be possible by incorporating AI technology in picture psychological analysis.

Keywords Image detection · Faster-RCNN · Computer vision · General image detection · DAP (Draw a Person)

J. W. Lee · D. Y. Kim
Department of IT Policy and Management, Graduate School, Soongsil University, Seoul, Korea
e-mail: dykim@ssu.ac.kr

J. H. Lee
J and Lee Co., Koung-gi, Korea
e-mail: haksa26@hanmail.net

G. Y. Gim (✉)
Department of Business Administration, Soongsil University, Seoul, Korea
e-mail: gygim@ssu.ac.kr

R. Lee (ed.), *Computer and Information Science 2021—Summer*,
Studies in Computational Intelligence 985,
https://doi.org/10.1007/978-3-030-79474-3_7

1 Introduction

Since 2012, computer vision technology has improved remarkably. Among them, with the development of deep learning, not only basic image classification in visual realization, but also object detection, object segmentation, image reconstruction, image synthesis, and other remarkable achievements have been achieved and are being used little by little in real life. There are various achievements in detecting objects in general images (e.g. cats, wolves, houses, people, objects) as well as detecting specific images (e.g. Eiffel Tower, Mona Lisa, Clinton). Furthermore, an attempt is being made to recognize the subdivided characteristics of an object. Detection the detailed characteristics of an object, such as Detection a person's motion or facial expression, can predict the expected next motion or emotion read from an expression. In the field of picture psychology diagnosis, it is said that experts find it difficult to find the picture features [1] in finding partial features of the picture and diagnosing the psychological state projected on the picture. It is thought that it is worth using not only in psychological analysis, but also in other industries if the feature area, which is a part of a person in the picture image, is automatically found through deep learning, like a live image.

Deep learning technique is conducted on areas that have psychological meaning in pictures of people. It examines whether the psychological state is predictable through partial images of the detected person and examines the possibility of contributing to psychological diagnosis.

2 Literature Review

2.1 Image Object Detection

A current status survey has been conducted on the data types, targets, and detection areas used for Object Detection since 2000, focusing on the related Object Detection survey (Table 1).

2.2 Draw A Person (DAP) Test

Draw A Person (DAP) was devised by Goodenough in 1926 as a type of projection test in which human unconscious mind is projected through drawing [18]. The DAP Test is aimed at children and adolescents, is a test that expresses unconscious psychological phenomena perceived toward oneself and others. It focuses on emotional behavior problems and intelligence tests [19]. After that, in 1956, Otomo set up 17 areas for each person's body for portrait painting and suggested appropriate analysis standards for personality diagnosis [20]. Table 2 is a part of the Otomo DAP Personality test

Table 1 Operational definition

No.	Survey title	Author (year)	Venue	Data type			Target		Area	
				Video	Photo	Picture	People	Others	All	Part
1	Detecting faces in images: a survey [2]	Yang et al. (2002)	PAMI		O		O			O
2	On road vehicle detection: a review [3]	Sun et al. (2006)	PAMI	O				O	O	
3	Monocular pedestrian detection: survey and experiments [4]	Enzweiler and Gavrila (2009)	PAMI		O		O	O	O	
4	Survey of pedestrian detection for advanced driver assistance systems [5]	Geronimo et al. (2010)	PAMI		O		O	O	O	O
5	Context based object categorization: a critical survey [6]	Galleguillos and Belongie (2010)	CVIU		O			O	O	
6	Pedestrian detection: an evaluation of the state of the art [7]	Dollar et al. (2012)	PAMI	O			O		O	
7	Object class detection: a survey [8]	Zhang et al. (2013)	ACM CS		O		O	O	O	
8	Representation learning: a review and new perspectives [9]	Bengio et al. (2013)	PAMI		O			O	O	
9	Salient object detection: a survey [10]	Borji et al. (2014)	arXiv		O			O	O	
10	A survey on face detection in the wild: past, present and future [11]	Zafeiriou et al. (2015)	CVIU		O		O			O
11	Text detection and recognition in imagery: a survey [12]	Ye and Doermann (2015)	PAMI		O			O	O	

(continued)

Table 1 (continued)

No.	Survey title	Author (year)	Venue	Data type			Target		Area	
				Video	Photo	Picture	People	Others	All	Part
12	Tutorial: tools for efficient object [13]	2015	ICCV15		O		O	O	O	O
13	A survey on deep learning in medical image analysis [14]	Litjens et al. (2017)	MIA	O			O			O
14	Recent advances in convolutional neural networks [15]	Gu et al. (2018)	PR			O		O	O	
15	Tutorial: deep learning for objects and scenes	2017	CVPR17		O			O	O	
16	Deep learning for generic object detection [16]	Ours (2019)	VISI		O			O	O	
17	The Open Images Dataset V4 [17]	Kuznetsova et al. (2020)	GoogleAI		O		O	O	O	O

Table 2 Otomo DAP Personality test analysis criterion (some examples)

Part	Personality analysis criterion		Male	Female
Head	So big. (More than 1/2 of the tors)	Aggressiveness by daydreaming, unconscious, intellectual satisfaction	S	S
	Small head	Obsessive–compulsive neurosis	Y	Y
	No hair	Schizophrenic	Sch	Sch
Eye	Emphasize the eyes	Perverted curiosity, hyperactive	Sch	Sch
	Closed eyes	Physical self-image	F	F
Hand	Clenched fist	Repressed aggression	F	F
	Turn your arm	Trying to suppress aggressive and hostile emotions	Sch	Sch
Cloth	Emphasize the button	Maternal dependence infantile maladjustment	F	F

analysis criterion. The signs of S, F, Sch, and F are the signs of biblical factors, and psychological treatment is required when the index score is 80–85 or higher [21].

Table 2 is a part of the Otomo DAP Personality test analysis criterion. The signs of S, F, Sch, and F are the signs of biblical factors, and psychological treatment is required when the index score is 80–85 or higher [21]. In the interpretation of the portrait painting test, a picture with a large head is related to activeness and unconsciousness by fantasy, and it also means satisfaction with intellectual things. Conversely, a small head is associated with obsessive compulsion or neurosis. In addition, when the hand is hidden, aggressive and hostile emotions are reflected, and the illustration of a fisted hand also reflects repressed aggression. Buttons on clothing, emphasizing pockets, etc. mean compensation for dependence and deficit.

2.3 Object Detection Algorithm

Object detection is a sub-technology of Computer Vision technology and is an automated technique that identifies objects of interest in an image by separating them from the background. Object recognition is largely classified into Classification, Object Detection, and Segmentation. Figure 1 shows (a) image classification aims to allocate the label of the class corresponding to the image category regardless of the position value of the image, and (b) object detection is the position of the object in the region of interest through bounding boxes. The purpose is to perform and classification. (c) Semantic segmentation aims to predict the label corresponding to the category for each pixel without distinguishing object instances, and (d) instance segmentation corresponds to a special setting of object detection, and at the pixel level (level) is used to distinguish it from other object instances [22].

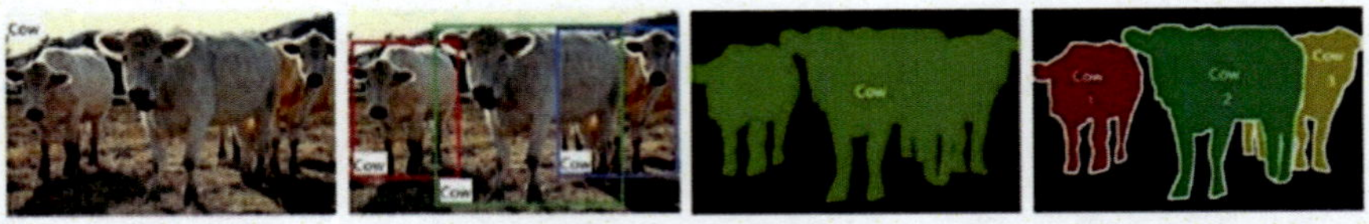

(a) Image Classification (b) Object Detection (c) Semantic Segmentation (d) Instance Segmentation

Fig. 1 Comparison of different visual recognition tasks in computer vision

Object Detection is basically able to hit multiple boxes even if there are multiple objects in one image. There are two main methods of detection algorithm. There are One-Stage Method and Two-Stage Method. The One-stage Method is a fast-integrated method that proposes the location of an object and handles classification at the same time. The Two-Stage Method has high accuracy by proposing and finding the area where the object is located, and then classifying the types of objects existing in the area [16].

Representative algorithms include YOLO [23], SSD [24], and YOLOv2 [25] for one-stage detection, and two-stage detection VGGNet, GoogLeNet, Fast RCNN [26], Faster RCNN [27], and Mask R-CNN [28], DnserNet [29], etc. According to the important value of the domain used in practice, algorithms with high speed and high accuracy should be selected and used, and algorithms with improved speed and performance are being studied.

3 Methodology

The procedure for detecting the partial region of a figure image is proposed as follows.

Step 1: Collect data

Build a data set by collecting the actual DAP (Drawing a Person) Test results. In accordance with the purpose of the thesis, drawing images are collected instead of actual images.

Step 2: Pretreatment step

As a pre-processing process, we define a label for the image area that will be psychologically meaningful. Mark the area and label the collected image.

In this paper, not a label definition method (class designation method) in general image detection, but a label for partial objects (e.g., emphasized eyes, emphasized hair, etc.) necessary for psychological analysis including personality, and a form containing behavior or state (Define the label as a hidden arm, a missing leg, etc.).

Step 3: Model training

It trains with Faster R-CNN method to generate a model, and finally performs verification.

In order to detect areas in which psychological characteristics are reflected and general areas in human images by referring to Otomo DAP Personality test analysis criterion. In addition, since the mental state must be predicted, the accuracy of the detection algorithm must be high. We intend to use Faster R-CNN, which is faster among the two-stage algorithms rather than the one-stage method.

Faster R-CNN is attracting attention because it can detect objects at a relatively high speed as an algorithm for detecting objects in images. That is, the method of extracting the proposal area was included in the entire neural network to reduce processing time and achieve consistency in learning. The network that detects the proposed region is called the RPN (resion proposal network), and Faster R-CNN integrates the process that existed separately into the CNN to share the CNN's network parameters, so that the entire process from input to output It operates as one network [30].

Step 4: Verification

At step 3, we check the learning rate of the algorithm with test data based on the learned model. At step 4, the re-learning process is repeated until the target values (mAP, loss, detection) are achieved.

4 Experiment and Verification

4.1 Experimental Method

This experiment uses data labeled for partial areas such as head, shoulders, hands, and attachments of clothing, and the deep learning object detection algorithm is trained with Faster R-CNN model to perform object detection experiments. The PC environment used is GPU GTX 1080 Ti, CPU inter i7-9080, RAM 16G DDR4.

4.1.1 Target Value

As there is no study on the learning rate of partial detection of hand-drawn person-picture, we will study faster R-CNNN up to 0.05 (5%) error loss and set a goal of 85% or higher of mAP by referring to the 2012 PASCAL VOC 2012 [31] test competition (Fig. 2).

	training data	mAP (%)	test time (sec/img)
Faster R-CNN [9]	07+12	76.4	0.42
Faster R-CNN +++ [9]	07+12+COCO	**85.6**	3.36
R-FCN	07+12	79.5	0.17
R-FCN multi-sc train	07+12	80.5	0.17
R-FCN multi-sc train	07+12+COCO	**83.6**	0.17

Fig. 2 Comparisons on PASCAL VOC 2012 test set using ResNet-101 [31]

4.1.2 Data

The data used in the experiment downloaded images from websites such as Google. Most of them were extremely normal, making it difficult to find images that could reveal psychological features. Therefore, based on the research paper related to the picture psychology test, the people's picture in case of a problem and the post-psychotherapy picture were collected together and the pictures of the normal and abnormal groups (those with psychological difficulties) of the data to be studied were collected in a ratio of 1:1. Some created images by performing actual Drawing a Person (DAP) tests and a total of 226 picture images were used for learning.

4.1.3 Labeling

In order to label the image, it was important how to set the label. If it is a task for general object detection, it can be set to 'person', 'head', 'face', 'foot', and 'hand'. Referring to Otomo's DAP Personality test analysis criterion [20], the characteristics of images that can find psychological factors in addition to the ambience of the body are also defined as labels. These features include, for example, 'to draw a large head', 'to omit a head', 'a fisted hand', 'to open arms', and 'to emphasize the eyes'. For these characteristics, labels are defined in the form of 'head_big', 'head_skip', 'hand_rock', 'arm_outer', and 'eye_stree'.

The image label work was done with the Python open library LabelImg [32]. The number of defined labels was 136, but the number of objects labeled after labeling image data was 105.

4.1.4 Training Model

In this experiment, an object detection experiment was conducted by learning with the Faster R-CNN model, which is a detection algorithm with high recognition accuracy, using labeling data for partial images showing psychological features. The code used in the experiment was learned by referring to Tensorflow model (Object detection) [33] API.

4.2 *Training*

4.2.1 Experiment 1

Among the 226 images of human figures, 192 pieces were selected for training data (85%) and 34 for test data (15%). The results of training using Faster R-CNN algorithm are shown in Table 3 (Fig. 3).

According to the first result, mAP (mean average precision) 83.6% means that the accuracy is 83.6% close when the detected area and the correct answer are compared, and the loss is a loss indicator, with a learning error range of 0–1, which is close to 0 The more it gets, the more it means there is nothing to learn. Usually a loss of about 0.05 can be seen as good learning. In the experiment, mAP was 83.6%, which did not achieve the target value (85%), but it can be confirmed that the hand-drawn human image is also detected as a partial object. When looking at the object detection rate, 92.5% of the object is basically human. In the partial area, the fisted hand (hand_lock) showed 92.0% and the head showed 97.0% detection rate. The average number of partial object detection in one picture is 6.2, which means that there are areas that were not detected in the picture. For example, in (a) the head was not detected, and in (b) the hand and leg were not detected. (c) In the figure, the shape of the pocket of the clothes was detected as a fisted hand. The rounded square is considered to be detected as it is similar to the hand expressed by a circle. Additional picture data was needed so that there was no detection error. The loss is 0.136, which means we need to proceed further.

4.2.2 Experiment 2

In the second experiment, data was increased to increase the accuracy rate. In order to augment the data of the first 226 pictures, the size of the pictures was reduced and enlarged, and the number of pictures of 226 was increased by three times, and deep learning was learned with 678 pictures. The learning results after resizing are shown in Table 4 (Fig. 4).

As a result of learning by doubling the amount of data, the number of objects detected per picture was 9.2, which is about 50% higher than the first experiment. Overall, the detection rate of partial objects was higher than in the previous experiment. Fingers that were not found in the first experiment were also identified in figure (b). Retraining was attempted with a loss of 0.126.

Table 3 1st training performance indicator

Person	Head	Head_big	Face	Mouth	Body	Shoulder	Arm_outer	Arm	Hand	Hand_lock	Leg	mAP	Loss	Detection
92.5	97.0	76.6	84.7	59.3	91.0	89.5	90.5	63.0	79.5	92.0	79.8	83.6	0.136	6.2

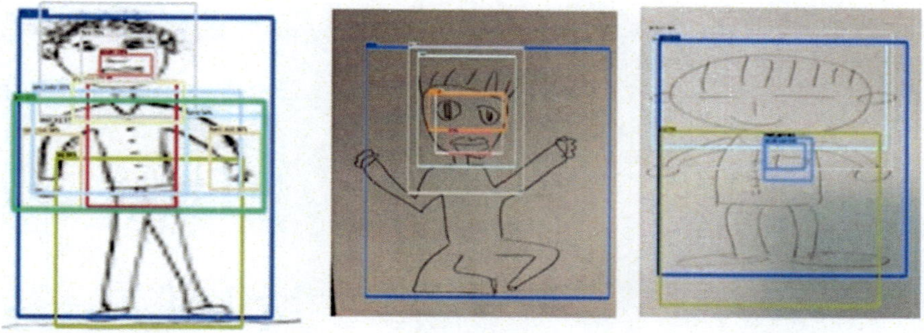

(a) Picture 1 (b) Picture 2 (c) Picture 3

Fig. 3 Partial detection result of hand drawn people picture

4.2.3 Experiment 3

Re-learning was conducted with the data that had been learned the second time. Several re-learnings were performed until the loss fell to the 0.05 level (Table 5 and Fig. 5).

As a result of the third experiment, re-learning was repeatedly performed. The best loss was 0.086. The detection accuracy of person, head, head_big, face, body, and finger was above 95%.

4.2.4 Experiment 4

When looking at the results of the 3rd experiment, it was difficult to expect a good learning rate with only the reduced or enlarged image of the picture, so the amount of data was increased by reversing the image for data augmentation. Deep learning training was conducted with the data increased to 2034 pages through vertical and horizontal reversal. The results are shown in Table 6 (Figs. 6 and 7).

As a result of the fourth experiment, the Loss decreased to 0.05 and the average accuracy was improved by 93.1%.

Buttons on clothes that were not found in previous experiments were detected in figures (a) and (c). Drawing clothing attachments is an important clue in psychological testing analysis. In the fourth experiment, images such as buttons where the amount of learning was not large were detected.

Table 4 2nd training performance indicator

Person	Head	Head_big	Face	Mouth	Body	Shoulder	Arm_outer	Arm	Hand	Hand_lock	Finger	Leg	mAP	Loss	Detection
98.7	98.4	79	90	76	96.5	93.3	88	86.4	88.5	89.3	84.2	87.4	89.2	0.126	9.2

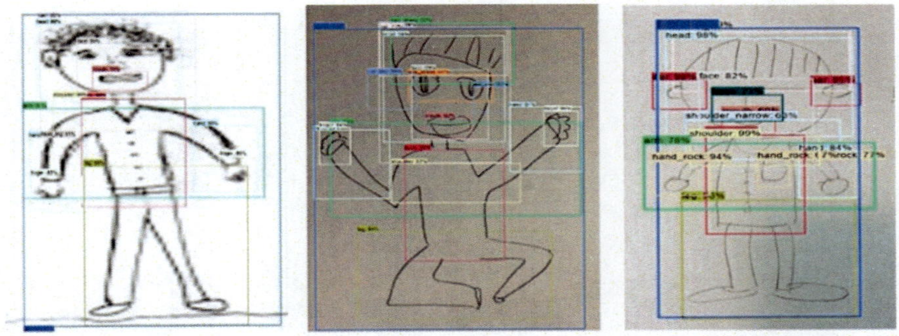

(a) Picture 1 (b) Picture 2 (c) Picture 3

Fig. 4 Partial detection result of hand drawn people picture

4.3 Verification

After the experiment up to the fourth round, in the case of other images that were not used in the experiment, the test was conducted with the algorithm obtained in the fourth experiment to see how the image part is detected (Fig. 8).

As a result of the test, in the case of (a) figure, mAP is 95.5%, in the case of (b) figure, mAP is 91.6%, indicating that the detection rate is above the target mAP of 85%.

4.4 Analysis

Based on the target mAp, loss, and detection, mAp was achieved in the second experiment, but the loss was not significantly different from the first time, so the loss fell through retraining. In the third experiment, the loss fell through retraining, but the loss did not fall below 0.08 by repeating the retraining alone. In addition to the basic type of image, additional types of images (e.g., images belonging to the basic image such as pockets, buttons, belts, etc. or additionally attached images) were not recognized, so the image was increased through image reversal and the fourth experiment was conducted I did. As a result of the 4th experiment, the loss dropped to 0.05, and an additional image of the button in the image to be searched was additionally detected (Table 7).

Table 5 3rd training performance indicator

Person	Head	Head_big	Face	Mouth	Body	Shoulder	Arm_outer	Arm	Hand	Hand_lock	Finger	Leg	mAP	Loss	Detection
98.7	95.8	87.3	98.7	74.7	96.5	97	–	94.6	94.3	93.5	95	94.3	92.3	0.086	10.5

(a) Picture 1 (b) Picture 2 (c) Picture 3

Fig. 5 Partial detection result of hand drawn people picture

5 Conclusions

5.1 Research Summary and Implication

This study examined partial detection of non-realistic images used in psychology.

In the case of existing real-life image detection, the use of non-realistic images (e.g., pictures) such as psychological tests are limited, so this paper has tested whether non-realistic images can be detected through AI. It was confirmed that detection is possible.

In particular, it was confirmed that detection is possible even for images that are not in basic form recognized as important for psychological analysis (e.g., images in the form of additional/attachment such as buttons, belts, pockets, etc.).

Through this, it was confirmed that AI technology can be used in DAP analysis, which is widely used in psychological analysis of children, and it is believed that it is possible to detect standardized results and rapid analysis during initial psychological analysis through AI.

In this paper, we analyzed the pictures drawn by children, and there are differences in the form and expression of pictures drawn by children and adults, so it is necessary to confirm whether detection is possible through additional experiments.

By all means it is expected that faster and higher-level psychological analysis will be possible by incorporating AI technology in picture psychological analysis.

Table 6 4th training performance indicator

Person	Head	Head_big	Face	Mouth	Body	Shoulder	Arm_outer	Arm	Hand	Hand_lock	Finger	Leg	Button	mAP
98.8	98.8	87.6	98	84.3	98.7	92.3	78.3	94.3	88.2	91	90.5	99	91.5	93.1

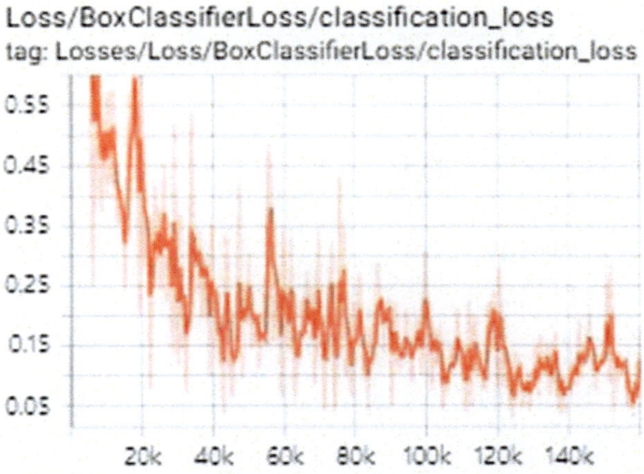

Loss
0.052

Fig. 6 Loss

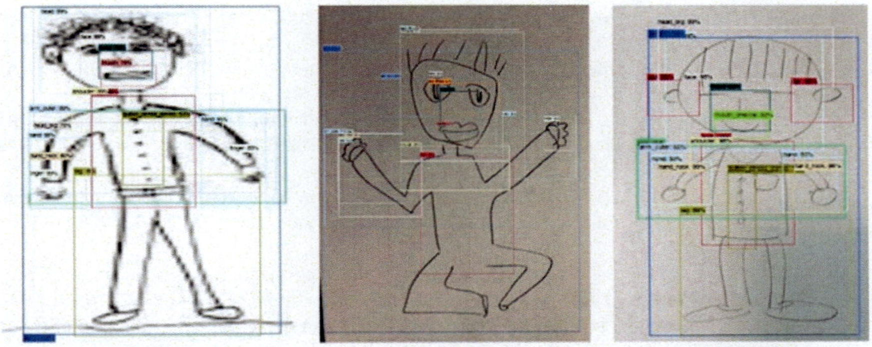

(a) Picture 1 (b) Picture 2 (c) Picture 3

Fig. 7 Partial detection result of hand drawn people picture

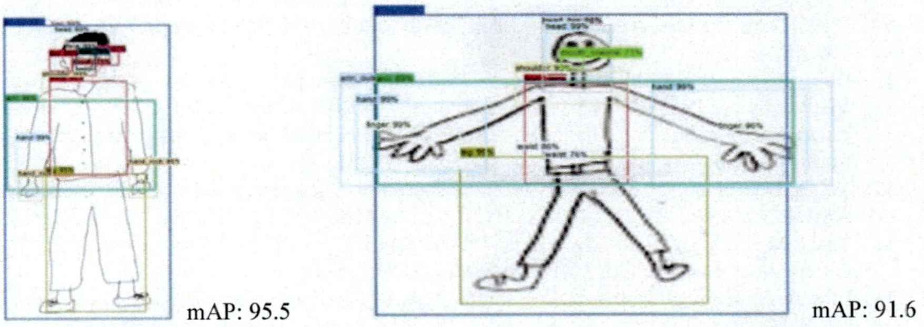

(a) Picture 1 (b) Picture 2

Fig. 8 Detection test result

Table 7 Deep learning training statistics

No	mAP (%)	Loss (%)	Detection count	Data count	Note
1	83.6	0.132	6.2	226	
2	89.2	0.126	9.2	678	Image resizing
3-1	90.3	0.107	9.7	678	Relearning
3-2	91.5	0.098	10	678	Relearning
3-3	92.1	0.087	10.1	678	Relearning
3-4	92.3	0.086	10.5	678	Relearning
4	93.1	0.052	11.7	2034	Image reversal

References

1. Kim, S., Yoo, S., Myung, R.H., Kim, S.: A framework of an expert system's knowledge for the diagnosis in art psychotherapy. J. Intell. Inf. Syst. **11**(1), 65–93 (2005)
2. Yang, M.-H., Kriegman, D.J., Ahuja, N.: Detecting faces in image: a survey—presentation. IEEE Trans. Pattern Anal. Mach. Intell. **24**(1), 1–25 (2002). https://citeseerx.ist.psu.edu/viewdoc/summary?doi=10.1.1.210.3372
3. Sun, Z., Bebis, G., Miller, R.: On-road vehicle detection: a review. IEEE Trans. Pattern Anal. Mach. Intell. **28**(5) (2006). https://doi.org/10.1109/TPAMI.2006.104
4. Enzweiler, M., Gavrila, D.M.: Monocular pedestrian detection: survey and experiments. IEEE Trans. Pattern Anal. Mach. Intell. **31**(12), 2179–2195 (2009). https://doi.org/10.1109/TPAMI.2008.260
5. Gerónimo, D., López, A.M., Sappa, A.D., Graf, T.: Survey of pedestrian detection for advanced driver assistance systems. IEEE Trans. Pattern Anal. Mach. Intell. **32**(7), 1239–1258 (2010). https://doi.org/10.1109/TPAMI.2009.122
6. Galleguillos, C., Belongie, S.: Context based object categorization: a critical survey. Comput. Vis. Image Underst. **114**(6), 712–722 (2010). https://doi.org/10.1016/j.cviu.2010.02.004
7. Dollár, P., Wojek, C., Schiele, B., Perona, P.: Pedestrian detection: an evaluation of the state of the art. IEEE Trans. Pattern Anal. Mach. Intell. **34**(4), 743–761 (2012). https://doi.org/10.1109/TPAMI.2011.155
8. Zhang, X., Yang, Y.H., Han, Z., Wang, H., Gao, C.: Object class detection: a survey. ACM Comput. Surv. **46**(1) (2013). https://doi.org/10.1145/2522968.2522978
9. Bengio, Y., Courville, A., Vincent, P.: Representation learning: a review and new perspectives. IEEE Trans. Pattern Anal. Mach. Intell. **35**(8), 1798–1828 (2013). https://doi.org/10.1109/TPAMI.2013.50
10. Borji, A., Cheng, M.M., Hou, Q., Jiang, H., Li, J.: Salient object detection: a survey. Comput. Vis. Media **5**(2), 117–150 (2019). https://doi.org/10.1007/s41095-019-0149-9
11. Zafeiriou, S., Zhang, C., Zhang, Z.: A survey on face detection in the wild: past, present and future. Comput. Vis. Image Underst. **138**, 1–24 (2015)
12. Ye, Q., Doermann, D.: Text detection and recognition in imagery: a survey. IEEE Trans. Pattern Anal. Mach. Intell. **37**(7), 1480–1500 (2015). https://doi.org/10.1109/TPAMI.2014.2366765
13. Yang, M.H., Kriegman, D.J., Ahuja, N.: Detecting faces in images: a survey. IEEE Trans. Pattern Anal. Mach. Intell. **24**(1), 34–58 (2002). https://doi.org/10.1109/34.982883
14. Litjens, G., Kooi, T., Bejnordi, B.E., Setio, A.A.A., Ciompi, F., Ghafoorian, M., van der Laak, J.A.W.M., van Ginneken, B., Sánchez, C.I.: A survey on deep learning in medical image analysis. Med. Image Anal. **42**, 60–88 (2017). https://doi.org/10.1016/j.media.2017.07.005
15. Zhang, Q., Zhang, M., Chen, T., Sun, Z., Ma, Y., Yu, B.: Recent advances in convolutional neural network acceleration. Neurocomputing **323**, 37–51 (2019). https://doi.org/10.1016/j.neucom.2018.09.038

16. Liu, L., Ouyang, W., Wang, X., Fieguth, P., Chen, J., Liu, X., Pietikäinen, M.: Deep learning for generic object detection: a survey. Int. J. Comput. Vis. **128**(2), 261–318 (2020). https://doi.org/10.1007/s11263-019-01247-4
17. Kuznetsova, A., Rom, H., Alldrin, N., Uijlings, J., Krasin, I., Pont-Tuset, J., Kamali, S., Popov, S., Malloci, M., Kolesnikov, A., Duerig, T., Ferrari, V.: The Open Images Dataset V4: unified image classification, object detection, and visual relationship detection at scale. Int. J. Comput. Vis. **128**(7), 1956–1981 (2020). https://doi.org/10.1007/s11263-020-01316-z
18. Rouse, M.J.: Children's drawings as measures of intellectual maturity. Stud. Art Educ. **6**(1), 49–51 (1964). https://doi.org/10.1080/00393541.1964.11650603
19. Joung, S., Park, Y.: Effect in emotion of children by personal color—focused on the best color. J. Korea Soc. Color Stud. **2014**(11), 25–38 (2014)
20. Kim, J.Y.: A study on the relationship between the defense mechanism type and the draw-a-person test. Master thesis, Hanyang University (2011)
21. Kim, D.C., Kim, S.S., Choi, Y.G.: Drawing a person. In: Picture and Psychological Diagnosis, p. 135. Yangseowon, Seoul (2014)
22. Wu, X., Sahoo, D., Hoi, S.C.H.: Recent advances in deep learning for object detection (2019). http://arxiv.org/abs/1908.03673
23. Redmon, J., Divvala, S., Girshick, R., Farhadi, A.: You only look once: unified, real-time object detection (2015). http://arxiv.org/abs/1506.02640
24. Liu, W., Anguelov, D., Erhan, D., Szegedy, C., Reed, S., Fu, C.-Y., Berg, A.C.: SSD: Single Shot MultiBox Detector (2015). https://doi.org/10.1007/978-3-319-46448-0_2
25. Redmon, J., Farhadi, A.: YOLO9000: better, faster, stronger (2016). https://arxiv.org/abs/1612.08242
26. Girshick, R.: Fast R-CNN (2015). https://arxiv.org/abs/1504.08083
27. Ren, S., He, K., Girshick, R., Sun, J.: Faster R-CNN: towards real-time object detection with region proposal networks (2015). http://arxiv.org/abs/1506.01497
28. He, K., Gkioxari, G., Dollár, P., Girshick, R.: Mask R-CNN (2017). http://arxiv.org/abs/1703.06870
29. Huang, G., Liu, Z., van der Maaten, L., Weinberger, K.Q.: Densely connected convolutional networks (2016). http://arxiv.org/abs/1608.06993
30. Lee, J.: A method of eye and lip region detection using faster R-CNN in face image. J. Korea Converg. Soc. **9**(8), 1–8 (2018)
31. Dai, J., Li, Y., He, K., Sun, J.: R-FCN: object detection via region-based fully convolutional networks. Adv. Neural Inf. Process. Syst. 379–387 (2016)
32. Tzutalin, LabelImg. Git Code. https://github.com/tzutalin/labelImg (2019)
33. Tensorflow Models. Git Code. https://github.com/tensorflow/models/blob/master/research/object_detection (2019)

Factors Affecting the Intention to Use Artificial Intelligence-Based Recruitment System: A Structural Equation Modeling (SEM) Approach

Jung Hee Lee, Ju Hyung Kim, Yong Hwan Kim, Yong Min Song, and Gwang Yong Gim

Abstract The controversy over fairness and objectivity in the job market, due to hiring irregularities, has led to calls for transparent and fair recruitment procedures. Advances in IT technology have led to the emergence of a non-face-to-face "AI recruitment system" in which artificial intelligence (AI) conducts interviews, instead of human interviews. As the introduction of the non-face-to-face method is encouraged in the hiring process due to the COVID-19 virus pandemics, the number of companies introducing AI recruitment systems is steadily increasing. In this study, the factors affecting the intention of use of AI-based recruitment system were analyzed by utilizing TOE and TAM. As a result, it was shown that the reliability, security, suitability, new technology, partiality, readiness, and legal and policy environment of the TOE affected the intention of using the system. It was also identified to have the moderating effect of the number of employees in the firm.

Keywords Artificial intelligence · AI recruitment system · AI interview job suitability · Recruitment procedure act

J. H. Lee · Y. M. Song
Department of Business Administration, Graduate School of Soongsil University, Seoul, Korea
e-mail: sym97dydals@naver.com

J. H. Kim · Y. H. Kim
Department of IT Policy and Management, Graduate School of Soongsil University, Seoul, Korea
e-mail: bmckorea@empas.com

Y. H. Kim
e-mail: berlnui@daum.net

G. Y. Gim (✉)
Department of Business Administration, Soongsil University, Seoul, Korea
e-mail: gygim@ssu.ac.kr

R. Lee (ed.), *Computer and Information Science 2021—Summer*,
Studies in Computational Intelligence 985,
https://doi.org/10.1007/978-3-030-79474-3_8

1 Introduction

The AI recruitment system reduces the time and cost of securing interview locations, provides recruitment services comparable to hiring professionals for small and medium-sized enterprises, and expands the scope of employees to be hired as applicants. As long as the Internet is connected to them regardless of the location, eliminating the room for human preferences of recruiters and increasing fairness [1]. For example, in the United States, they use programs such as IBM's Watson to identify personality and values through resume and social media analysis [2].

In 2020, a non-face-to-face approach was particularly recommended in recruitment due to the Pandemic of the COVID-19 virus, and the AI recruitment system began to be introduced in the hiring process. If the AI recruitment system can be included in the existing recruitment process and the non-face-to-face recruitment method can be implemented, it is necessary to examine the factors affecting intention to use of AI recruitment system. To proceed with the research, we analyze which factors affect the intention to use of the AI recruitment system using TAM and TOE.

2 Theoretical Background

2.1 *Artificial Intelligence and Recruitment*

Artificial intelligence (AI) is currently being used in various industries and the origin of the concept is quite old. At the 1956 American Conference of Dartmouth, cognitive scientists proposed the term "artificial intelligence," meaning "science and engineering making intelligent machines" [2]. Artificial intelligence is divided into strong and weak artificial intelligence depending on the autonomy of cognitive abilities [3]. In the past, "weak artificial intelligence" was the mainstream to replace repetitive tasks, but in recent years, consumers are increasingly interested in "strong artificial intelligence" technology as AI technology has been applied to various areas and consumers demand a higher level of intelligence [4].

This technological advance of AI constantly changes the company's recruitment strategy, such as the application selection process. AI based recruitment can be an effective process of human resource management [5]. In the U.S., 50 of Fortune 500 companies, including Unilever and Hilton Hotels, implemented the recruitment process using artificial intelligence [6].

2.2 *Recruitment Procedure Act in Korea*

Act on the Fairness of State Laws, abbreviated as the Recruitment Procedure Act, is a law that bans unfair solicitation, pressure, and coercion in hiring human resources

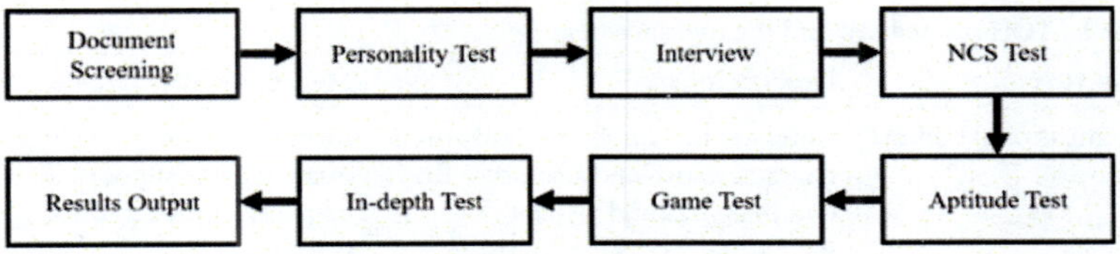

Fig. 1 Process of AI-based recruitment system (7 phase example)

and prohibits job seekers from requesting personal information, such as physical conditions unrelated to job performance.

A work-seeker shall not require a work-seeker to enter information not necessary for the performance of his/her duties in the application form, resume, or a letter of self-introduction, or collect evidence. Following personal information is prohibited from collecting and requesting: physical condition such as appearance, height, and weight, area of origin, marital status, and lineal ascendants of property, and educational background, occupation, and property of brothers and sisters. However, even if personal information is prohibited by law, collection and request can be made if necessary for the performance of duties.

2.3 AI Recruitment System

AI recruitment system is a system that can identify job suitability of applicants by using artificial intelligence and machine learning, and in the entire process of recruitment, the AI image analysis system uses facial recognition technology to analyze empathy, anomalies, confidence, positivity, criticality, curiosity, concentration, and sociality, and the AI voice analysis system detects even the smallest sounds around and analyzes the frequency of use of positive words, negative words, neutral words, and words. If the existing recruitment process was usually done through documents (i.e. resume, applications) screening, written test, and interview, the AI interview process depends on the needs of the company, but this study will give an example of the seven-step recruitment process. The process is shown in Fig. 1.

2.4 Technology, Organization, and Environment (TOE) Framework

The TOE framework proposed by Tornatzky et al. [7] describes the factors influencing the organization's adoption of information technology from three perspectives: technological context, organizational context, and environmental context. The summary of TOE framework is shown in Table 1 [8].

Table 1 TOE framework and the measurement

Context	Description	Measurement
Organizational context	Refers to the unique organizational characteristics and resources that a company has and affects technology adoption by itself	Firm size, firm scope, top management support, staff capacity/competence, surplus resources
Technological context	It is important to determine whether technology can be introduced and if the technology currently being applied is suitable for the firm	Relative advantage, compatibility, complexity, trialability, perceived barriers, IT/IS infrastructure
Environmental context	Means the business activities of a company and includes the industry, competitors, government regulations, business partners, etc. to which the company belongs	Competitive pressure, environmental uncertainty, government incentives and regulations

2.5 Technology Acceptance Model (TAM)

Davis et al. [9]'s Technology Acceptance Model is a proposed model for predicting users' acceptance in the introduction of new technologies and modifies the proposed reasonable theory of behavior in Fishbein and Ajzen to explain and predict people's behavior [10]. The perceived usefulness that the TAM model expects to improve work performance by using a new information system and the perceived ease of use that it considers easy to use a new information system are used in conjunction with the intention to accept the technology. While traditional technology acceptance models include attitude variables as parameters, many prior researches have shown that attitude are not important parameters [11].

3 Research Design

The recruitment industry is one of the latest service industries where the convergence of existing NCS based blind hiring regulations and IT is evolving into AI recruitment systems. The system is an integrated form of 4th industrial technology. To identify the factors affecting the intention to use of AI recruitment system in the organization, this research designed the research model which is presented in Fig. 2. Also, the research examined moderating effect of the number of employees in the firm.

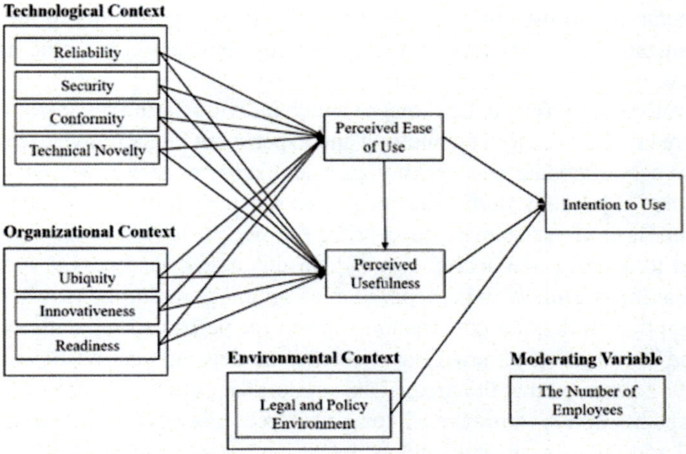

Fig. 2 Research model

3.1 Research Hypothesis

Reliability suggests in previous studies that people have different degrees of trust in other people or objects, are considered individual characteristics, and that services have a positive relationship to the use and acceptance of new technologies [12]. In this study, reliability is defined to the extent that AI recruitment system are believed to be reliable.

Security means the degree to which new approaches are prevented from hacking attacks in the information technology area, and the degree to which authentication errors are blocked [13]. In this study, security was defined as the degree of belief that the use of AI recruitment systems would be safe without personal information infringement or leakage.

Tornatzky and Klein [14] conducted a study that comprehensively compares previous prior studies, explaining that conformity, along with relative advantages and complexity, is a major influence on innovation acceptance. In this study, conformity was defined to the extent that the recruitment system is appropriate for the organization.

Technical novelty was adopted as a construct to determine the effect of the novelty on acceptance within AI recruitment systems where research is not actively conducted. In this study, we defined technical novelty as the degree of novelty that users feel about the technology that AI recruitment systems have.

Kannan et al. [15] addressed that the ubiquity enables customers to reach customers regardless of location in real time, receive diverse information regardless of their location, and enable real-time communication. Meanwhile, Ku and An [16] found that ubiquitous convenience affects perceived ease of use and that there is no significant impact on perceived usefulness in the study of factors affecting the acceptability of u-Commerce. This study defined the ubiquity to the extent that the organization believes that AI recruitment system services can be applied well anywhere.

Innovativeness refers to the extent to which individuals and organizations are able to explore new information technology and experiment on their own to exercise their power to accommodate new technologies and services relatively faster than other consumers [17]. In this study, innovativeness was used as the same measurement.

Readiness is a variable for measuring the degree to which an organization is prepared to adopt a new technology [8]. In this study, readiness is defined to the extent that the organization is prepared to adopt an AI recruitment system.

Dasgupta [18] defined government support for personnel regulations as encouraging the introduction of new IT systems for the development of the company. Teo et al. [19] confirmed that the greater the support for policy, the greater the intention of the organization to introduce a human resource information system. In this study, the legal and policy environment was defined as extent to which law and policy are important for the introduction of AI recruitment systems.

Perceived ease of use is a factor that directly affects perceived usefulness, and the easier and more convenient users perceive the new technology or system to be, the more useful it is [20]. Accordingly, this study defined perceived ease of use to the extent that user feel easy to use the systems. Perceived usefulness refers to the degree to which users believe that the use of information systems can increase performance and is useful in their daily lives [9]. Bhattacherjee [21] explained that perceived usefulness had a significant effect on the intention to use. This research defined usefulness to the extent that it is believed to be beneficial for the purpose of use.

Lastly, a hypothesis was established with the number of employees as a moderating variable to find that the number of employees in the firm makes difference in the relationship of the factors. The summary of the hypothesis is presented in Table 2.

4 Empirical Analysis

To collect the data for the analysis, an online survey was conducted to measure the variables. The survey targets include personnel department workers who understand the artificial intelligence recruitment system or decision makers of corporation in the team of management, planning and strategy. Of the total 216 survey responses, 187 data were used for the final analysis, except for those that were missing or insincere response. A frequency analysis was conducted to examine the demographic characteristics of the respondents, and the results were shown in Table 3. For statistical analysis of all process, the SPSS and AMOS tools were used.

Table 2 Summary of the research hypothesis

Hypothesis
H1: Reliability of AI recruitment systems has a positive impact on perceived ease of use
H2: Reliability of AI recruitment systems has a positive impact on perceived usefulness
H3: Security of AI recruitment systems has a positive impact on perceived ease of use
H4: Security of AI recruitment systems has a positive impact on perceived usefulness
H5: Conformity of AI recruitment systems has a positive impact on perceived ease of use
H6: Conformity of AI recruitment systems has a positive impact on perceived usefulness
H7: Technical novelty of AI recruitment systems has a positive impact on perceived ease of use
H8: Technical novelty of AI recruitment systems has a positive impact on perceived usefulness
H9: Ubiquity of AI recruitment systems has a positive impact on perceived ease of use
H10: Ubiquity of AI recruitment systems has a positive impact on perceived usefulness
H11: Innovativeness of AI recruitment systems has a positive impact on perceived ease of use
H12: Innovativeness of AI recruitment systems has a positive impact on perceived usefulness
H13: Readiness of AI recruitment systems has a positive impact on perceived ease of use
H14: Readiness of AI recruitment systems has a positive impact on perceived usefulness
H15: Legal and policy env. of AI recruitment systems has a positive impact on perceived usefulness
H16: Perceived ease of use of AI recruitment systems has a positive impact on the perceived usefulness
H17: Perceived ease of use of AI recruitment systems has a positive impact on the intention to use
H18: Perceived usefulness of AI recruitment systems has a positive impact on the intention to use
H19: The number of employees in the company (less than 100 employees, more than 100 employees) moderate the relationship of factors affecting the intention to use of the AI recruitment system

4.1 *Reliability, Convergence Validity, and Discriminant Validity*

Table 4 show the values of factor loading, Cronbach's alpha, composite reliability (CR), and average variance extracted (AVE) for all constructs. Reliability assessments indicate the stability and consistency of instrument's measurement in a research model, and the reliability of the measurement is commonly evaluated through Cronbach's Alpha. If the Cronbach's Alpha is 0.7 or higher, it can be evaluated that internal consistency of the extracted constructs has been obtained [22], and the Cronbach's Alpha values of all the structures has exceeded the criteria.

For the convergence validity assessment, CR and AVE were used to evaluate the assessment. All the values of CR are satisfied the recommended 0.7 criteria. Also AVE for each construct are acceptable by exceeding the threshold 0.5 [23].

Correlations and AVE were used to test the discriminant validity. AVE should be higher than the square of the highest value of the correlations [23]. The diagonal

Table 3 Demographic characteristics of survey respondents

Section		Freq.	%
Individual			
Work area	Personnel	39	20.9
	Planning/Strategy	45	24.1
	Accounting/Finance	39	20.9
	R&D	19	10.2
	Others	45	24.1
Position	Clerk	40	21.4
	(Assistant) Manager	48	25.7
	(Deputy) General Manager	35	18.7
	Director and above	53	28.3
	Others	11	5.9
Industry	Finance	12	6.4
	Public institution	23	12.3
	IT	31	16.6
	Manufacturing	16	8.6
	Distribution	19	10.2
	Service	39	20.9
	Others	47	25.1
Number of employees in their firm (person)	1–100	110	58.8
	+100	77	41.2

values in Table 5 represent AVE for each constructs, off-diagonal values indicate the correlations between each latent variable. The highest correlation is 0.851 and the squared value is 0.724. AVE for each construct was higher than the threshold of 0.724, so the discriminant validity was satisfied.

4.2 Goodness of Fit

This assessed the measurement model by conducting a CFA. Through the analysis of measurement model, goodness of fit indexes for the model are extracted and they can estimate how well the model represents the data. The values of indexes with recommended criteria are shown in Table 6. While the GFI and RMR were below the recommended levels, the overall model fit was considered to be fine because other indices were significantly higher or lower than the threshold [24–26].

Table 4 Construct reliability and convergent validity

Constructs	Indicator	Factor loading	Cronbach's α	CR	AVE
Ubiquity	UB1	0.831	0.912	0.902	0.756
	UB2	0.806			
	UB3	0.755			
Innovativeness	IV1	0.875	0.932	0.888	0.726
	IV2	0.833			
	IV3	0.783			
Readiness	RD1	0.829	0.928	0.898	0.745
	RD2	0.816			
	RD3	0.779			
Reliability	RA1	0.783	0.957	0.939	0.756
	RA2	0.763			
	RA3	0.755			
	RA4	0.739			
	RA5	0.735			
Security	SC1	0.829	0.938	0.905	0.761
	SC2	0.824			
	SC3	0.783			
Conformity	CF1	0.684	0.938	0.920	0.741
	CF2	0.651			
	CF3	0.639			
	CF4	0.548			
Technical novelty	TN1	0.800	0.900	0.887	0.723
	TN2	0.784			
	TN3	0.717			
Legal and policy environment	LPE1	0.775	0.931	0.899	0.749
	LPE2	0.712			
	LPE3	0.693			
Perceived ease of use	PEU1	0.801	0.959	0.945	0.810
	PEU2	0.787			
	PEU3	0.784			
	PEU4	0.778			
Perceived usefulness	PU1	0.850	0.964	0.933	0.777
	PU2	0.824			
	PU3	0.820			
	PU4	0.774			
Intention to use	IU1	830	0.967	0.942	0.765

(continued)

Table 4 (continued)

Constructs	Indicator	Factor loading	Cronbach's α	CR	AVE
	IU2	0.816			
	IU3	0.814			
	IU4	0.792			
	IU5	0.751			

Table 5 Construct correlations and discriminant validity

	UB	IV	RD	RA	SC	CF	TN	LPE	PEU	PU	IU
UB	0.756										
IV	0.51	0.726									
RD	0.423	0.511	0.745								
RA	0.692	0.643	0.532	0.756							
SC	0.536	0.539	0.487	0.739	0.761						
CF	0.693	0.633	0.661	0.851	0.713	0.741					
TN	0.612	0.483	0.730	0.544	0.469	0.629	0.723				
LPE	0.562	0.633	0.719	0.671	0.516	0.703	0.764	0.749			
PEU	0.744	0.604	0.583	0.843	0.732	0.842	0.668	0.774	0.81		
PU	0.689	0.627	0.595	0.795	0.694	0.799	0.536	0.644	0.797	0.777	
IU	0.614	0.606	0.529	0.745	0.654	0.761	0.533	0.72	0.782	0.826	0.765

Table 6 Goodness of fit indexes for the model

Indexes	Result value	Recommend
Chi-square/degree of freedom	1.434	<3
Goodness of fit index (GFI)	0.802	>0.9
Normed fit index (NFI)	0.902	>0.9
Comparative fit index (CFI)	0.968	>0.9
Incremental fit index (IFI)	0.968	>0.9
Root mean square error of approximation (RMSEA)	0.048	<0.08
Root mean square residual (RMR)	0.051	<0.05

4.3 Structural Model for Hypotheses Testing

The tests for each hypothesis in the research model were conducted using structural equation modelling via AMOS. The path analysis results are shown in Table 7, and the path coefficients are listed as standardized coefficients.

As a result of the analysis in the organizational context, only the ubiquity and readiness have a positive impact on the perceived ease of use. This means that laying

Table 7 Summary of hypothesis test results

Hypothesis	Path coefficient	p	Result
H1: Reliability → perceived ease of use	0.243	0.001	Supported
H2: Reliability → perceived usefulness	0.107	0.103	Rejected
H3: Security → perceived ease of use	0.1	0.098	Rejected
H4: Security → perceived usefulness	−0.012	0.81	Rejected
H5: Conformity → perceived ease of use	0.2	0.015	Supported
H6: Conformity → perceived usefulness	−0.088	0.207	Rejected
H7: Technical novelty → perceived ease of use	0.243	0.018	Supported
H8: Technical novelty → perceived usefulness	0.28	0.001	Supported
H9: Ubiquity → perceived ease of use	0.15	0.029	Supported
H10: Ubiquity → perceived usefulness	0.133	0.021	Supported
H11: Innovativeness → perceived ease of use	0.215	0.053	Rejected
H12: Innovativeness → perceived usefulness	0.22	0.018	Supported
H13: Readiness → perceived ease of use	−0.146	0.081	Rejected
H14: Readiness → perceived usefulness	0.247	***	Supported
H15: Legal and policy environment → intention to use	0.232	***	Supported
H16: Perceived ease of use → perceived usefulness	0.159	0.032	Supported
H17: Perceived ease of use → intention to use	0.533	***	Supported
H18: Perceived usefulness → intention to use	0.185	0.021	Supported

the foundation for adopting the AI-based recruitment system from an organization's perspective makes it easier to use the system. In contrast, the variables in technical context had a significant impact on perceived ease of use and usefulness, except for the two paths. This address that at the period when the AI-based recruitment system has yet to be actively utilized, the automation of recruitment process requires confidence in technical trust and security, and acceptable technical novelty and system suitability are considered as important point for the performance. The recruitment procedures are generally regulated by the government's law and policy for the fairness of the process and selection. In the results, Law and policy environment in the environmental context had a direct effect on intention to use. In the path associated with the TAM constructs, all the relations have a significant effect. In particular, the coefficient between the path of ease of use and intention to use was 0.533, which indicates that convenience of application is most important attributes to the workers at the corporates.

Table 8 Summary of multi-group analysis (PLS-MGA)

Relationship	Difference	*p*	Number of employees			
			1–100		+100	
			β	*p*	β	*p*
H2: RA → PU	−0.319*	0.037	0.129	0.177	0.447	***
H15: LPE → IU	0.374*	0.022	0.341	***	-0.032	0.81
H17: PEU → IU	0.386*	0.027	0.622	***	0.236	0.121
H18: PU → IU	−0.555**	0.004	0.043	0.614	0.599	***

Note: *p<.05, **p<.01, ***p<.001

4.4 Moderating Effect Analysis

A multi-group analysis (PLS-MGA) was conducted using smartPLS 3.3.2 as the last step of the empirical analysis. The groups were divided into two groups with under 100 employees group and more than 100 employees group. First, we confirmed the possibility of a moderating effect analysis through the Measurement Invariance Assessment (MICOM) and conducted a PLS-MGA. The results suggested that there were differences only for the four paths as shown in Table 8.

5 Conclusions

In this study, the factors that affect the intention to use of the artificial intelligence recruitment system were examined. The social environment changes and the rapid development of the IT environment caused by the COVID-19 around the world have brought many changes in the recruitment culture as well. Moreover, the government has revised the Recruitment Procedure Act, as controversy mounts over the fairness of hiring irregularities day after day.

In addition, at a time when there is a growing debate among job seekers about the fairness of the recruitment process, research has been conducted. But the research for the view of work-seeker is still in the beginning stage and should be studied after the implementation process. For job seekers, negative and positive public opinion exists together because AI systems that apply artificial intelligence technology are not fully generalized and lack understanding. In this study, we conducted the research only from the perspective of the company and conducted a survey of all departments' workers, but in future research, the researches to conduct a survey of the company's personnel managers and job seekers to study from other perspectives are needed, and the social impact or effectiveness in terms of the AI recruitment system adoption.

References

1. Park, S.H.: Everything About AI Competency Testing. EDUCE, Seoul, Korea (2020)
2. Lee, H.W., Lee, R.S., Jung, C.G.: The impact of artificial intelligence adoption in candidates screening and job interview on intentions to apply. J. Inf. Syst. **28**(2), 25–52 (2019)
3. Hengstler, M., Enkel, E., Duelli, S.: Applied artificial intelligence and trust—the case of autonomous vehicles and medical assistance devices. Technol. Forecast. Soc. Chang. **105**, 105–120 (2016)
4. Bae, W.Y., Maeng, S.G., Han, H.S.: Artificial intelligence is expanding into reality. KEIT PD Issue Report 2016-09. Korea Evaluation Institute of Industrial Technology (2016). https://www.bioin.or.kr/board.do?num=264308&cmd=view&bid=tech
5. Van Esch, P., Black, J.S., Ferolie, J.: Marketing AI recruitment: the next phase in job application and selection. Comput. Hum. Behav. **90**, 215–222 (2019)
6. Lee, S.W.: AI interviewers appear in U.S. corporate recruitment... Concerns about fairness in evaluation. Seoul (2018). http://m.seoul.co.kr/news/newsView.php?id=20180921500157&cp=seoul#csidx8724d62a156885fb0eaafdef95a2697
7. Tornatzky, L.G., Fleischer, M., Chakrabarti, A.K.: Processes of Technological Innovation. Lexington Books (1990)
8. Gangwar, H., Date, H., Raoot, A.D.: Review on IT adoption: insights from recent technologies. J. Enterp. Inf. Manage. (2014)
9. Davis, F.D., Bagozzi, R.P., Warshaw, P.R.: User acceptance of computer technology: a comparison of two theoretical models. Manage. Sci. **35**(8), 982–1003 (1989)
10. Legris, P., Ingham, J., Collerette, P.: Why do people use information technology? A critical review of the technology acceptance model. Inf. Manage. **40**(3), 191–204 (2003)
11. Venkatesh, V., Davis, F.D.: A theoretical extension of the technology acceptance model: four longitudinal field studies. Manage. Sci. **46**(2), 186–204 (2000)
12. Lee, H.Y., Ahn, H., Han, I.: VCR: virtual community recommender using the technology acceptance model and the user's needs type. Expert Syst. Appl. **33**(4), 984–995 (2007)
13. Schierz, P.G., Schilke, O., Wirtz, B.W.: Understanding consumer acceptance of mobile payment services: an empirical analysis. Electron. Commer. Res. Appl. **9**(3), 209–216 (2010)
14. Tornatzky, L.G., Klein, K.J.: Innovation characteristics and innovation adoption-implementation: a meta-analysis of findings. IEEE Trans. Eng. Manage. **1**, 28–45 (1982)
15. Kannan, P.K., Chang, A.M., Whinston, A.B.: Wireless commerce: marketing issues and possibilities. In: Proceedings of the 34th Annual Hawaii International Conference on System Sciences, 6 pp. IEEE (2001)
16. Ku, K.B., An, B.J.: An empirical study on important factors of ubiquitous commerce acceptance. Int. Account. Res. **18**, 1–21 (2007)
17. Rogers, E.M., Shoemaker, F.F.: Communication of innovations: a cross-cultural approach (1971)
18. Dasgupta, S.: Supporting ideas with data. Econ. Pol. Wkly **34**(50), 3513–3515 (1999)
19. Teo, T.S., Lim, G.S., Fedric, S.A.: The adoption and diffusion of human resources information systems in Singapore. Asia Pac. J. Hum. Resour. **45**(1), 44–62 (2007)
20. Chin, W.W., Todd, P.A.: On the use, usefulness, and ease of use of structural equation modeling in MIS research: a note of caution. MIS Q. 237–246 (1995)
21. Bhattacherjee, A.: Understanding information systems continuance: an expectation-confirmation model. MIS Q. 351–370 (2001)
22. Nunnally, J.C., Berstein, I.H.: Psychometric Theory. McGraw-Hill, New York (1994)

23. Fornell, C., Larcker, D.F.: Evaluating structural equation models with unobservable variables and measurement error. J. Mark. Res. **18**(1), 39–50 (1981)
24. Taylor, S., Todd, P.A.: Understanding information technology usage: a test of competing models. Inf. Syst. Res. **6**(2), 144–176 (1995)
25. Gefen, D., Straub, D., Boudreau, M.C.: Structural equation modeling and regression: guidelines for research practice. Commun. Assoc. Inf. Syst. **4**(1), 7 (2000)
26. Cangur, S., Ercan, I.: Comparison of model fit indices used in structural equation modeling under multivariate normality. J. Mod. Appl. Stat. Methods **14**(1), 14 (2015)

A Comparative Study of Vectorization Approaches for Detecting Inconsistent Method Names

Tomoya Minehisa, Hirohisa Aman, Tomoyuki Yokogawa, and Minoru Kawahara

Abstract Methods (functions) are the fundamental components of the software. Programmers usually grasp a method's behavior by looking at the method's name. Hence, the name of a method should be a summary of what the method does. There has been a study utilizing Word2Vec, Doc2Vec, and the convolutional neural network (CNN) to evaluate the consistency between a method's name and body in an automated way. While the conventional evaluation procedure detects inconsistent method names successfully, its CNN-based vectorization model construction requires costly computations. This paper focuses on such a computational cost and proposes to replace it with another lightweight vectorization approach. The comparative study using four alternative approaches proved that the Sent2Vec, one of the alternatives, can build the vectorization model 14 times faster than the conventional one while maintaining the capability of detecting inconsistent method names.

Keywords Method name · Inconsistency · Vectorization · Automated assessment

1 Introduction

A software system consists of various objects or modules, and each object or module has several methods (functions) to implement its functionality. Methods are units

T. Minehisa
Graduate School of Science and Engineering, Ehime University, Matsuyama, Ehime 790-8577, Japan
e-mail: mine@se.cite.ehime-u.ac.jp

H. Aman (✉) · M. Kawahara
Center for Information Technology, Ehime University, Matsuyama, Ehime 790-8577, Japan
e-mail: aman@ehime-u.ac.jp

M. Kawahara
e-mail: kawahara@ehime-u.ac.jp

T. Yokogawa
Faculty of Computer Science and Systems Engineering, Okayama Prefectural University, Soja, Okayama 719-1197, Japan
e-mail: t-yokoga@cse.oka-pu.ac.jp

R. Lee (ed.), *Computer and Information Science 2021—Summer*,
Studies in Computational Intelligence 985,
https://doi.org/10.1007/978-3-030-79474-3_9

of programs that perform specific tasks and play as the fundamental components of the software. When programmers code methods, they have to describe not only the bodies of methods but also the names of them. Although a method's name works as the method's identifier, it is also expected to give valuable information about its behavior [1, 2]. That is, a method's name should also be a summary of what the method does.

If a method's name does not represent the method's behavior properly, such an inconsistency adversely affects the readability of source code [3, 4] and would prevent effective software quality management. Although it is hard to find the single absolutely-correct way of naming methods, some naming conventions have been widely used in the programming world [5–8]. For instance, it is common to name Java methods in the *camelCase* style using English words or abbreviations.

However, such a naming standard gives a necessary condition of well-named methods but not sufficient condition. For example, `getName` is a common name consisting of two English words concatenated in the camelCase style. While programmers would expect that the method returns a name property of the object to which the method belongs, the name does not guarantee that the method behaves as expected. That is, even if a method's name confirms a standard naming convention, we need to perform a code review to ensure the consistency between a method's name and body (behavior). Indeed, the code review is one of the most worthwhile activities to find inconsistent (or inappropriate) names [9]. On the other hand, the manual code review is also a costly activity because it involves one or more reviewers other than the method's author. Thus, it is beneficial to assist the manual code review by automatically detecting inconsistent method names while considering the method bodies.

Recently, Liu et al. [10] proposed a promising automated procedure for detecting inconsistent method names, utilizing natural language processing techniques and machine learning techniques. Liu et al. employed Word2Vec [11], Doc2Vec [12], and the convolutional neural network (CNN) [13] to vectorize methods' names and bodies. Such a vectorization enables us to measure the similarity between methods quantitatively. When evaluating a method's name, the procedure searches other methods having similar bodies without looking at their names. If the method under evaluation has an *inconsistent* name, it would be *dissimilar* to any similar-body methods' names. Their automated procedure successfully examines the consistency between a method's name and body based on that notion. However, it also has a challenge regarding the computational cost, i.e., it tends to require a long time to build the CNN-based vectorization model. That challenge motivated us to use a lightweight alternative vectorization approach for detecting inconsistent method names.

This paper considers four alternative approaches and examines their practicality from model construction time and judgment performance perspectives through a comparative empirical study. The key contributions of this paper are as follows:

- This paper proposes to replace the vectorization approach used in the previous work [10] with a lighter-weight alternative.

- The comparative study using the same dataset [14] as the previous work proved that Sent2Vec [15], one of the alternatives, can build the vectorization model 14 times faster than the conventional one while maintaining the capability of detecting inconsistent method names.

This paper is structured as follows. Section 2 briefly outlines the way of naming methods and its evaluation in terms of the consistency between a method's name and body. Section 3 describes the conventional automated procedure for detecting inconsistent method names and its challenge regarding computational cost, then proposes lightweight alternatives for the costly vectorization part. Section 4 reports the results of our comparative empirical study and discusses the practicality of the alternative approaches. Finally, our conclusion and future work are described in Sect. 5.

2 Method Name and Its Evaluation

In this section, we describe the common trend of naming methods. Then, we focus on the consistency between a method's name and body and consider its evaluation.

2.1 Method Name

Programmers must give a name to each method when they declare or define it. The method name itself or the combination of the name and the arguments serves as the method's identifier. Programmers may give any name to a method as long as it is compliant with the grammar of the programming language that they use. Nonetheless, a method's name should summarize the method's behavior toward better program comprehension and quality management [3, 4]. To this end, there have been standard naming conventions [5–8], and programmers have performed their programming activities under such conventions.

For example, the following three features are common to the standard naming conventions for Java.

1. A method name comprises one or more English words or abbreviations.
2. A method name is organized in the *camelCase* style when it contains two or more words (terms).
3. A method name starts with a verb or an auxiliary verb.

It is the standard way of naming methods using English words or abbreviations to express the methods' behaviors properly. Programmers often concatenate two or more words (terms) to make a method name more descriptive. Such a compounded name looks like a phrase explaining what the method does. In Java programs, it is general to form a compounded method name by the camelCase style [5]. Moreover, the first (initial) term forming a method name is usually a verb or an auxiliary verb. The initial verb represents the fundamental action done by the method. For example, when a method name starts with "get" such as `getPath`, programmers expect the method returns the object corresponding to the remaining part of the name (e.g., `Path`). When a method name starts with a *be* verb or an auxiliary verb, it looks like a yes-no question such as `isReady` and `canWrite`. Programmers expect the method returns the Boolean value, which is the answer to the yes-no question corresponding to the method name.

2.2 The Consistency Between the Method's Name and Body

As mentioned above, a method's name should properly represent its behavior, i.e., the source code described in the method body. For example, let us consider a method returning the "`title`" attribute of a dialog object (a GUI component). Most programmers would name the method "`getTitle`." On the other hand, if the method name were "`setTitle`," it would mislead many programmers into expecting that the method sets the value of the "`title`" attribute. While the former name (`getTitle`) is consistent with the method body, the latter one (`setTitle`) is an *inconsistent* method name.

Although the above example is extreme, there are inconsistent method names even in popular open-source software products. For instance, the Apache Ant ver.1.7.0 had an inconsistent method name, "`isCaching`" [16]. Because the name represents a question, programmers expect the method returns the Boolean value answering that question. However, the method was designed to assign its argument (parameter) to the attribute "`caching`." Hence, it is an inconsistent method name even though the naming style is entirely compliant with the above three standard features of Java method names (see Sect. 2.1) and looks like a well-chosen name. We would fail to notice the inconsistency unless we review the source code written in the method body.

2.3 Method Name Consistency Evaluation

To evaluate the consistency between a method's name and body, we need well-defined criteria regarding the method name. Although there is no absolutely-correct way of naming a method, we have several standard ways of it for Java, mentioned in Sect. 2.1, including

1. the use of English words or abbreviations,
2. the term concatenation in the camelCase style, and
3. the use of a verb or an auxiliary verb at the beginning of the name.

However, the last feature has some exceptions in the real. For example, we often see `toString`, `valueOf`, and `iterator` as Java method names while they do not have the above third feature. Nonetheless, the word (term) appearing at the beginning of the method name plays a significant role in representing the method's behavior. For instance, a method whose name is `toString` would produce a string representation of the object and return it. The initial word `to` is a simple representation of such a conversion of the data type to String. Thus, to accept the above examples as well, we will revise the third feature as:

3'. the use of a keyword (key term) representing the behavior, at the beginning of the name.

Here, we consider how to evaluate a method name in an automated manner. We can automate the checking by preparing an English dictionary on the first feature, except for abbreviations. Although it is not easy to prepare an exhaustive abbreviation dictionary [17], we can mitigate the issue using an English abbreviation website[1] or abbreviation expansion studies [18–21]. For the second feature (i.e., camelCase), we can also perform an automated check of a method name using a simple pattern matching operation. On the other hand, checking the third feature we have just revised above is challenging. To judge whether the initial word (term) of a method name appropriately represents the method's behavior or not, we need to review the source code written in the method body. Because the code review is costly and not an easy-to-perform task [22, 23], it is challenging but demanded to automate the above judgment toward an efficient evaluation of the method name consistency. The automatic detection of inconsistent method names would promote the efficiency of code review and quality management.

3 Vectorization-Based Detection of Inconsistent Method Names

Lately, to detect inconsistent method names, Liu et al. [10] proposed an automated procedure using the vectorization approaches, including the Word2Vec, the Doc2Vec, and the convolutional neural network (CNN). We give outlines of their automated procedure together with the vectorization approaches in Sects. 3.1 and 3.2. Then, we discuss its challenge regarding the computational cost in Sect. 3.3.

3.1 Vectorization Approaches Used in the Conventional Evaluation Procedure

We briefly describe the three vectorization approaches before introducing the automated procedure by Liu et al. (the conventional one).

[1] https://www.abbreviations.com/abbreviation/.

3.1.1 Word2Vec

The Word2Vec [11] is a sophisticated approach for producing distributed representations of words (word embeddings) by learning text data using the neural network. The produced word embeddings are multi-dimensional vector representations of words (word vectors) while considering the word semantics. Using the word vectors, we can quantitatively evaluate the semantic similarity between words. There are two major learning models for Word2Vec: the continuous bag-of-words (CBoW) and the skip-gram.

The CBoW model trains the neural network to predict a word using its surrounding words. For example, we consider a training sentence, "He drinks a glass of juice every morning." We mask a word in the sentence as "He drinks a glass of [ ] every morning" and train the neural network to predict the word corresponding to the blank from the corpus of text. The neural network tunes the weight coefficients between its neurons through training with many different sentences. Once the training is finished, the weight coefficients are used as the set of word vectors. Although the correct word for the blank in the above example is "juice," we can get a consistent sentence by putting another word into the blank such as "water" and "milk." Hence, the weight coefficients, i.e., the word vectors corresponding to "juice," "water," and "milk," may be similar to each other. It means that those words may have similarities in semantics, and the Word2Vec successfully represents the semantic similarity among words in a quantitative manner.

The skip-gram model, another learning model, uses the training data in a similar but opposite direction to CBoW. It trains the neural network to predict the surrounding words (e.g., "glass," "of," "every," "morning") from the focused word (e.g., "juice"). The tuned neural network can produce word vectors as well as that learned by CBoW.

3.1.2 Doc2Vec

The Doc2Vec [12] is an extension of the Word2Vec to vectorize documents, i.e., paragraphs consist of two or more words. The fundamental mechanism of Doc2Vec is common to Word2Vec, and Doc2Vec uses not only sentences but also document IDs as its training data. There are two learning models: the distributed memory model of paragraph vector (PV-DM) and the distributed bag of words version of paragraph vector (PV-DBoW). PV-DM and PV-DBoW are based on CBoW and skip-gram of Word2Vec, respectively. Although both the learning models tune the Doc2Vec neural networks, PV-DBoW does not consider the order of words in the paragraph.

3.1.3 CNN

The convolutional neural network (CNN) [13] is a deep neural network to deal with two-dimensional data like an image, which has several convolutional layers and pooling layers. A convolutional layer uses a data filter (a smaller window) to focus only on

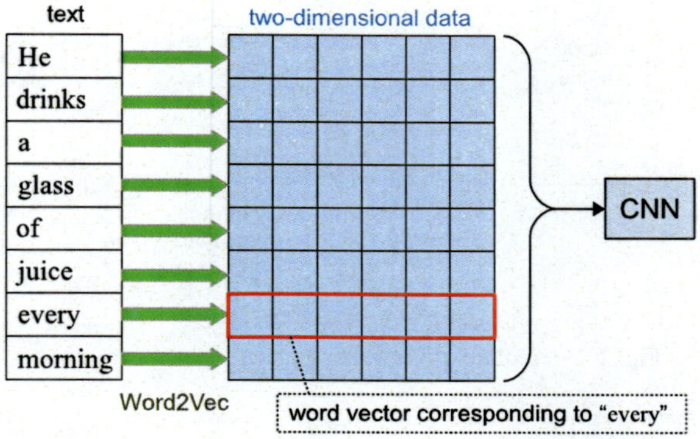

Fig. 1 An example illustrating the conversion of text data to two-dimensional data for CNN

a portion of the input data and computes a focused portion's representative feature. Then it convolves the input data into a collection of the computed representative features (a feature map) by moving the filter over the input data. A pooling layer condenses a feature map into a smaller-sized map to enhance the robustness against fluctuations of data and reduce the computational cost. A CNN processes the input data by alternately connecting convolutional layers and pooling ones, and it has been known as an advanced feature extractor in image recognition. By using a word vectorization method like Word2Vec, we can convert a document into two-dimensional data like an image (Fig. 1). Thus, because we can extract a document's features through CNN, it can also be an approach for vectorizing the document (producing the vector of the document's features).

3.2 Conventional Procedure

In this subsection, we explain the vectorization-based conventional procedure [10] for detecting inconsistent method names. The procedure consists of two parts: the vectorization part and the evaluation one.

3.2.1 Vectorization Part

Given a method, the conventional procedure vectorizes the method's name and its body separately (Fig. 2). Method names are vectorized using Doc2Vec, and method bodies are done using Word2Vec and CNN (Word2Vec + CNN), respectively. We describe more details of the vectorizations below.

Given a method name, we split it into words (terms) according to the *camelCase* or the *snake_case*. For example, we split `getTitle` into `get` and `title`. The obtained sequence of words looks like a short phrase corresponding to the original

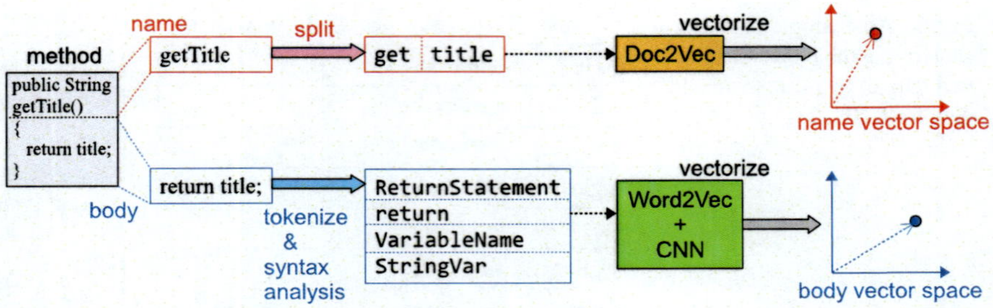

Fig. 2 An outline of vectorizing a method's name and body

method name, and we vectorize it using Doc2Vec (see the upper half of Fig. 2); In [10], the learning model of Doc2Vec is PV-DM.

For a method body, we perform the lexical analysis and the syntax analysis to produce the sequence of tokens except for delimiters. In the token sequence, we append the syntax label to each token, representing the token type in the abstract syntax tree. Furthermore, we anonymize all identifiers (e.g., variable names) appearing in the method body by replacing them with their type information. The identifier anonymization eliminates the potential bias caused by identifiers (names) and makes the token sequence express the program structure only. For example, when a method's body consists of the return statement "return title;" and "title" is a String variable, the corresponding token sequence is "`ReturnStatement`, `return`, `VariableName`, `StringVar`," where the first and third tokens give the syntax information. The remaining tokens are the keyword and the anonymized variable name, respectively. Given such a token sequence, Liu et al. proposed to vectorize it using Word2Vec + CNN mentioned in Sect. 3.1.3 (see the lower half of Fig. 2); The learning model of Word2Vec is skip-gram.

Notice that the two-dimensional data inputted to CNN must be fixed-size data, i.e., any data must have the same width and the same height. As we explained in Fig. 1, the data width and height correspond to the word vector's length and the number of tokens, respectively. Although we can easily prepare the word vectors with the same length by using the same Word2Vec model, the number of tokens depends on the method body of interest. Hence, Liu et al. empirically defined the sufficient number of tokens (=94) to express any method bodies in their dataset and fixed the data height. When the number of tokens is less than the predefined height, the remaining blank part is filled with zero vectors.

3.2.2 Evaluation Part

Next, we outline the evaluation part for judging whether a method's name is inconsistent or not. Let m_0 be the method of interest for the sake of convenience.

Firstly, we search for k methods that are the most similar to m_0 in the name vector space; k is given as the experimental parameter later. Although these k methods have similar names to m_0, we do not look at their bodies at all. Let M_{name} be the set of those k methods' names.

Secondly, we also explore k methods that are the most similar to m_0 in the *body* vector space. Let M_{body} be the set of those k methods' names. In contrast to M_{name}, the names included in M_{body} are the methods' names, whose bodies are similar to m_0 regardless of their names.

Although both M_{name} and M_{body} are the name sets of methods similar to m_0, their focuses are independent of each other: M_{name} focuses only on the name similarity, and M_{body} does only on the body similarity. Then, we look at the intersection (overlap) between M_{name} and M_{body}. If m_0's name is consistent with its body, there may be other methods with a similar name and a similar body. That is, M_{name} and M_{body} may overlap even though we built them separately. On the other hand, when m_0's name is inconsistent, the methods whose bodies are similar to m_0 would not have similar names.

Therefore, the conventional procedure judges that m_0's name is *inconsistent* if $M_{name} \cap M_{body} = \emptyset$. Notice that the name evaluation is performed by focusing only on the initial word (term) in the method name. The initial word of a method name primarily represents the method's behavior, as mentioned in Sect. 2. The remaining words (terms) usually give more details specific to the method or the object. However, they may become less important because the method's details are anonymized in the vectorization. For example, suppose m_0's name is `getTitle`, i.e., its initial word is `get`, and $M_{name} =$ { `get`, `to`, `set` }, where we show only the initial words of the method names in M_{name}. If $M_{body} =$ { `get`, `to`, `print` }, we consider m_0's name to be consistent. On the other hand, if $M_{body} =$ { `set`, `put`, `save` }, we regard m_0's name as an inconsistent method name because $M_{name} \cap M_{body} = \emptyset$.

3.3 Challenge of Conventional Procedure and Alternatives

We can automatically evaluate the method name consistency using the conventional procedure [10]. However, the conventional procedure has a challenge regarding computational cost. We performed a reproductive experiment of [10] and confirmed that the model construction requires much time and memory; we took over six hours to finish the model construction using an average personal computer.[2] Although the conventional procedure is a beneficial automated way of detecting inconsistent method names, its high computational cost may degrade the practicality when we want to keep the evaluation system up-to-date toward better detection performance. It is ideal for preparing the inconsistent name detection model in a lighter-weight manner to support manual code review.

[2]CPU: Intel Core i5-6600 3.3GHz; Memory: 16GB; OS: Linux 4.19.128.

Table 1 Vectorization approaches for method body

	Approach
Conventional approach	Word2Vec+CNN
Alternative approach (1)	fastText+CNN
Alternative approach (2)	Doc2Vec_DM
Alternative approach (3)	Doc2Vec_DBoW
Alternative approach (4)	Sent2Vec

Through our reproductive experiment, we realized that the vectorization model construction using Word2Vec+CNN was a highly time-consuming part. Thus, we considered replacing that part with another lighter-weight vectorization approach, and this is our motivation for the comparative study presented in this paper. If an alternative approach can build the vectorization model faster than the conventional Word2Vec+CNN while maintaining the inconsistent name detection performance, the alternative approach makes the conventional procedure more practical. We propose four alternatives shown in Table 1. We selected these alternatives because we can implement all of them using the same Python environment and library; we considered such homogeneity is essential to compare the approaches fairly. We briefly describe them below.

3.3.1 fastText+CNN

First, we prepare an alternative approach by replacing Word2Vec with the fastText [24] while using CNN; We denote it by "fastText+CNN." The fastText is also an approach for vectorizing words, and it can finish its learning faster than Word2Vec (skip-gram). Furthermore, fastText has an advantage in the vectorization of unlearned words using the notion of "subword." The subwords of a word are portions of the word; For example, the subwords of "apple" are "ap," "pp," "pl," and "le." To produce a word vector, the fastText uses both the learned words and the learned subwords. Hence, it may deal with even an unlearned word using the subwords matched to the unlearned word.

3.3.2 Doc2Vec_DM

Next, we consider a straightforward alternative which is the same approach as the method name vectorization. That is, we adopt Doc2Vec, whose learning model is PV-DM; we denote it by "Doc2Vec_DM."

3.3.3 Doc2Vec_DBoW

We can use another learning model for Doc2Vec, PV-DBoW. Thus, we append it to our alternative approaches; we denote the alternative by "Doc2Vec_DBoW."

3.3.4 Sent2Vec

Sent2Vec [15] is yet another approach for vectorizing documents. It is also based on Word2Vec. Although Doc2Vec uses the words appearing in the document and the document's ID, Sent2Vec focuses on the words and the N-grams of words. After completing the learning, Sent2Vec produces the document vector by averaging all of the word vectors and the N-gram vectors. Because Sent2Vec is also a promising vectorization approach, we use it as an alternative in our comparative study.

4 Comparative Study

We have described the conventional vectorization-based procedure for detecting inconsistent method names and its challenge of computational cost. To solve the challenge, we have considered replacing the time-consuming part with a lighter-weight alternative. The part of interest is the vectorization model (Word2Vec+CNN) construction for method bodies in the conventional procedure. Our alternatives include four vectorization approaches shown in Table 1, and our idea is to use one of them instead of Word2Vec+CNN. We conducted a comparative study to examine the usefulness of our idea and the above alternatives, and we report the results in this section.

4.1 Aim

This study aims at showing the usefulness of the above alternative vectorization approaches (see Table 1) through a quantitative comparative study. We evaluate the usefulness from two perspectives, including the computational (model construction) time and the detection accuracy. Suppose an alternative approach could prepare the evaluation system much faster than the conventional one while maintaining the same detection accuracy level. In that case, the alternative produces a more practical procedure for detecting inconsistent method names.

4.2 Dataset and Computational Environment

We use the same dataset as the previous work [10], which is available from Liu et al.'s website [14]. The dataset consists of 2,119,218 Java methods (pairs of method name and body) from 430 open-source software projects, and the set is divided into the training dataset and the test dataset (see Table 2). We use the training dataset to build the vectorization models, and we utilize the test dataset to evaluate the accuracy of detecting inconsistent method names. The test dataset includes the correct answers of method name consistency that were decided by the hand.

Table 2 Number of Java methods included in the dataset

Data type	Number of methods
Training	2,116,413
Rest	2805

Table 3 Computational environment

Item	Description
CPU	Intel Core i5-6600 3.3 GHz
Memory	16 GB
OS	Ubuntu 20.04.1 (Linux 4.19.128) on Windows 10 Education
Software	Python 3.8.5, gensim 3.8.3, keras 2.4.3, numpy 1.18.5, fastText 0.9.2, Sent2vec 1.0

Table 3 shows our computational environment. Because one of our research purposes is to examine the practicality of the studied approaches, we use an average personal computer in this study.

4.3 Procedure

We performed our comparative study in the following steps.

1. Vectorization model construction
   We need to vectorize the method names and bodies included in the training dataset to construct the vector spaces for evaluation. We use Doc2Vec_DM to vectorize method names according to the conventional procedure (see Fig. 2). On the other hand, for the vectorization of method bodies, we use the conventional approach (Word2Vec+CNN) and the above four alternatives independently. Hence, we prepare different five vectorization models for representing method bodies. Table 4 presents the hyperparameters we adopted in this study, according to the previous work [10]. Although the above alternative vectorization approaches were not used in the previous study, we decided their parameters so that they have the same vector size (dimension) as the conventional one; We adopt the libraries' default values for the other parameters that are not specified in the table.
   For each approach, we measure the time to complete the vectorization model construction using all training data.
2. Method name evaluation
   For each method included in the test data, we vectorize its name and body and explore the closest *k* methods in the vector spaces of the training data. According

Table 4 Hyperparameters used in the vectorization approaches for method bodies

Approach	Parameters
Word2Vec	Size = 300, min_count = 1, window = 4, sg = 1, alpha = 1e–2
CNN	input_shape = (94, 300, 1), units = 1000, Activation ="softmax" (for the output), Activation ="relu" (for the others), Loss ="mean absolute error", optimizer ="SGD"
Doc2Vec_DM	vector_size = 1000, min_count = 1, window = 94, dm = 0, alpha = 0.025
Doc2Vec_DBoW	vector_size = 1000, min_count = 1, window = 94, dm = 1, alpha = 0.025
Sent2Vec	minCount = 1, lr = 0.025, dim = 1000, ws = 5, wordNgrams = 2
fastText	minCount = 1, lr = 1e–2, dim = 300, ws = 4

Table 5 Time to complete the vectorization model construction using all training data

	Approach	Computational time	
Conventional	Word2Vec+CNN	22,855 s	(6 h 21 min)
Alternative (1)	fastText+CNN	18,011 s	(5 h)
Alternative (2)	Doc2Vec_DM	2093 s	(35 min)
Alternative (3)	Doc2Vec_DBoW	2058 s	(34 min)
Alternative (4)	Sent2Vec	**1579 s**	(**26 min**)

to the previous work [10], we adopt different six values: $k = 1, 5, 10, 20, 30,$ and 40. Then, we decide whether the method name is inconsistent or not, as we explained in Sect. 3.2.2. Because the test dataset contains the correct answers regarding the name consistency, we can evaluate the judgment quantitatively. The evaluation criteria include the precision, recall, and F value (the harmonic mean between the precision and the recall).

We compare the conventional approach with the alternative ones using the vectorization model construction time measured in step 1 and the evaluation scores obtained in step 2.

4.4 Results

We report the results of our comparison in this subsection.

First, we show the results regarding the model construction time. Table 5 presents each approach's time to build the vectorization model using all training data, which we explained in step 1 of Sect. 4.3. In the table, we highlight the best time (shortest time) in boldface. The conventional approach and the alternative (1) took five or more hours to complete the model constructions. On the other hand, the fastest approach is the alternative (4) using Sent2Vec. When we use Sent2Vec as the method body vectorization approach, we can prepare the model for evaluation in 1579 s (about 26 min); It is about 14 times faster than the conventional approach.

Next, we present the performance evaluations of the studied approaches in detecting inconsistent method names. Table 6 shows the precision, recall, and F value of each vectorization approach for $k = 1, 5, 10, 20, 30$, and 40. In the table, we emphasize

Table 6 Performance evaluations of each approach in detecting inconsistent method names

		Criteria		
k	Approach	Precision	Recall	F value
$k = 1$	Word2Vec+CNN	**0.529**	**0.851**	**0.653**
	fastText+CNN	0.536	0.833	0.652
	Doc2Vec_DM	0.491	0.840	0.620
	Doc2Vec_DBoW	0.484	0.703	0.573
	Sent2Vec	0.525	0.840	0.646
$k = 5$	Word2Vec+CNN	0.544	0.604	0.573
	fastText+CNN	**0.546**	0.599	0.572
	Doc2Vec_DM	0.504	0.517	0.510
	Doc2Vec_DBoW	0.464	0.445	0.454
	Sent2Vec	0.540	**0.624**	**0.579**
$k = 10$	Word2Vec+CNN	**0.557**	0.470	0.510
	fastText+CNN	0.553	0.455	0.499
	Doc2Vec_DM	0.502	0.313	0.385
	Doc2Vec_DBoW	0.474	0.326	0.386
	Sent2Vec	0.550	**0.481**	**0.513**
$k = 20$	Word2Vec+CNN	0.533	**0.679**	**0.597**
	fastText+CNN	0.568	0.233	0.330
	Doc2Vec_DM	0.499	0.211	0.297
	Doc2Vec_DBoW	0.482	0.242	0.323
	Sent2Vec	**0.570**	0.242	0.340
$k = 30$	Word2Vec+CNN	0.558	0.172	0.263
	fastText+CNN	0.545	0.168	0.257
	Doc2Vec_DM	0.481	0.153	0.232
	Doc2Vec_DBoW	0.480	0.182	0.264
	Sent2Vec	**0.565**	**0.183**	**0.277**
$k = 40$	Word2Vec+CNN	0.562	0.146	0.232
	fastText+CNN	0.545	0.145	0.23
	Doc2Vec_DM	0.497	0.132	0.208
	Doc2Vec_DBoW	0.500	0.160	0.243
	Sent2Vec	**0.569**	**0.158**	**0.248**

the best performance value within each of the three criteria for each k, by boldface style.

As a result, the conventional approach (Word2Vec+CNN) showed the best performance in terms of all three criteria only when $k = 1$; When we set other values to k, other alternative approaches worked better for detecting inconsistent method names in terms of one or more criteria. Moreover, even when $k = 1$, the conventional approach did not stand out from the others, and Sent2Vec showed good performances close to the conventional one. Therefore, we can say that the conventional approach does not always the best way of vectorizing method bodies, and we may replace it with another approach without the loss of usefulness.

4.5 Discussion

We focused on the vectorization approach for method bodies in the inconsistent method name detection. Then, we compared the conventional approach, i.e., Word2Vec+CNN, with the four alternatives shown in Table 1 from the two perspectives: the computational cost required for the model construction and the inconsistent name detection performance.

On the model construction cost (computational time), while the conventional approach took 22,855 s (about 6.3 h), Sent2Vec could finish it in 1579 s (about 26 min) that corresponds to about 7% of the conventional one (see Fig. 3). We could also build two Doc2Vec models much faster than the conventional one, and their times were 34–35 min (about 9% of the conventional one). On the other hand, although the construction of fastText+CNN could also finish faster than Word2Vec+CNN, it took

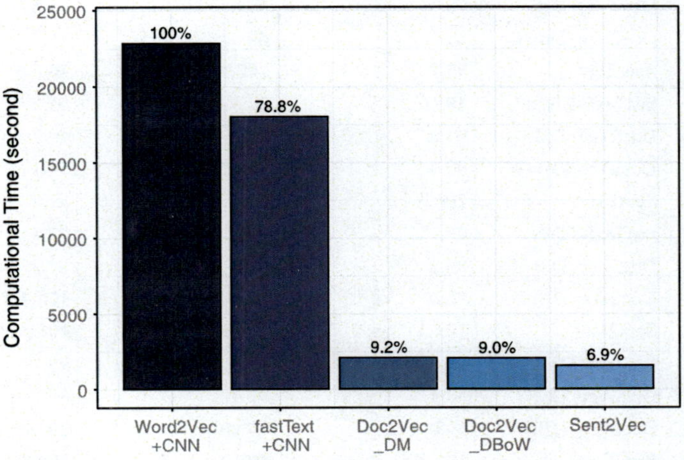

Fig. 3 A comparison of model construction times among approaches when we take Word2Vec+CNN's time as 100%

about five hours and did not seem to solve the challenge regarding the computational cost. Because both Wor2Vec+CNN and fastText+CNN took much longer than the remaining alternatives, we can reaffirm that CNN is the significant factor causing the computational cost challenge. Therefore, we can say Sent2Vec or Doc2Vec are better options for solving that challenge.

On the performance in detecting inconsistent method names, we proved that the conventional approach (Word2Vec+CNN) is not always the best way of vectorizing method bodies. By considering the above results on the computational time, we will focus on whether Doc2Vec or Sent2Vec can perform as well as Word2Vec+CNN. If Doc2Vec or Sent2Vec has almost the same performance evaluations as Word2Vec+CNN, they can be practical alternatives to the conventional approach. To compare their practicalities quantitatively, we consider the relative evaluation values of each alternative approach. Table 7 shows relative evaluation

Table 7 Relative evaluations of each alternative approach

		Relative values when we take Word2Vec+CNN's evaluation as 100		
k	Approach	Precision	Recall	F value
$k = 1$	fastText+CNN	**101.3**	97.9	**99.8**
	Doc2Vec_DM	92.8	**98.7**	94.9
	Doc2Vec_DBoW	91.5	82.6	87.7
	Sent2Vec	**99.2**	**98.7**	**98.9**
$k = 5$	fastText+CNN	**100.4**	**99.2**	**99.8**
	Doc2Vec_DM	92.6	85.6	89.0
	Doc2Vec_DBoW	85.3	73.7	79.2
	Sent2Vec	**99.3**	**103.3**	**101.0**
$k = 10$	fastText+CNN	**99.3**	96.8	97.8
	Doc2Vec_DM	90.1	66.6	75.5
	Doc2Vec_DBoW	85.1	69.4	75.7
	Sent2Vec	**98.7**	**102.3**	**100.6**
$k = 20$	fastText+CNN	**106.6**	34.3	55.3
	Doc2Vec_DM	93.6	31.1	49.7
	Doc2Vec_DBoW	90.4	35.6	54.1
	Sent2Vec	**106.9**	35.6	57.0
$k = 30$	fastText+CNN	97.7	97.7	97.7
	Doc2Vec_DM	86.2	89.0	88.2
	Doc2Vec_DBoW	86.0	**105.8**	**100.4**
	Sent2Vec	**101.3**	**106.4**	**105.3**
$k = 40$	fastText+CNN	97.0	**99.3**	**99.1**
	Doc2Vec_DM	88.4	90.4	89.7
	Doc2Vec_DBoW	89.0	**109.6**	**104.7**
	Sent2Vec	**101.2**	**108.2**	**106.9**

values when we take the conventional approach's evaluation value as 100. For example, because the precision values of Word2Vec+CNN and Sent2Vec when $k = 1$ are 0.529 and 0.525, respectively (see Table 6), the relative precision evaluation of Sent2Vec is 99.2 (=0.525/0.529×100). We consider that an alternative approach has almost the same performance as Word2Vec+CNN if the relative evaluation values' difference is within the margin of plus or minus 2%. Table 7 emphasizes the relative evaluations, whose values are equal to or greater than 98.0, by boldface. That is, the boldfaced values signify almost the same or better evaluations than the conventional approach.

From Table 7, we can see Sent2Vec's performance is almost equal to or better than that of the conventional approach (Word2Vec+CNN) in many cases; the relative performance values are shown by boldface in 16 out of 18 cases. Although we could build two Doc2Vec models much faster than the conventional approach, they showed lower performances in many cases (see non-boldface numbers), as shown in Table 7. While Sent2Vec is similar to Doc2Vec, Sent2Vec utilizes the notion of N-gram as well. Such a difference might work better for capturing the features of a method body and for replacing Word2Vec+CNN.

Thus, we can replace Word2Vec+CNN with Sent2Vec without the loss of performance in detecting inconsistent Java method names. Furthermore, we proved that Sent2Vec could be prepared about 14 times faster than the conventional approach; Sent2Vec's construction time was about 7% of the conventional one's time. Therefore, we conclude that Sent2Vec is a practical and lightweight alternative for Word2Vec+CNN in the inconsistent method name detection procedure. That is, by replacing Word2Vec+CNN with Sent2Vec in the conventional procedure, we can prepare the model for detecting inconsistent method names 14 times faster while maintaining the detection precision, recall, and F value.

Notice that our focus is on the preparation cost of the evaluation system and the system's evaluation performance (the detection precision, recall, and F value). Once the vectorization models were constructed, each model's speed to make judgments is at the same level as the others. Hence, replacing Word2Vec+CNN with Sent2Vec has an advantage in the situation that we want to keep the evaluation system up-to-date toward better automatic evaluation.

4.6 Threats to Validity

We describe our threats to validity below.

Construct Validity

We evaluated each approach's model construction time using the elapsed time that we measured as the difference between the construction start and its end. Thus, we could not eliminate the impact of other processes running in the operating system. However, because we did not run any other applications during the experiment, we consider that such an impact would be minor and did not significantly change the measured computational time.

Internal Validity

In general, the studied vectorization approaches' performances depend on their hyperparameter settings. However, we adopted the parameters according to the previous work [10] without tuning parameters because parameter tuning for each approach can also become another threat to internal validity. We further study the impacts of hyperparameters on our results as our future work.

We implemented all evaluation systems (including the conventional approach and all alternative ones) using the same Python environment to compare them fairly. However, the previous work used another library, DL4J.[3] Such a difference in implementation seems to affect the experimental results. We plan to replicate our study using DL4J and examine the difference in the future.

External Validity

Although we used Liu et al.'s dataset [14] in this study, those method data are limited to Java. Thus, we may not generalize our findings to the software written in other programming languages. It can be a threat to external validity in our study.

5 Conclusion and Future Work

We have focused on the consistency between a method's name and body in this paper. Although the primary way of evaluating the method name consistency had been the manual code review in the past, there has been a novel automated procedure [10] using natural language techniques and machine learning techniques recently. The conventional automated procedure vectorizes the method's name and body independently, and it evaluates the similarity among the feature vectors to detect an inconsistent method name. Although the conventional procedure successfully detects inconsistent method names, it has a challenge regarding the computational cost: it takes a long time to prepare the vectorization model.

Toward a lighter-weight evaluation, we considered four alternative vectorization approaches: (1) fastText+CNN, (2) Doc2Vec_DM, (3) Doc2Vec_DBoW, and (4) Sent2Vec. Then, we conducted a comparative study of the conventional approach (Word2Vec+CNN) and the above four alternative ones. As a result, we proved that Sent2Vec could be a practical lightweight alternative approach for vectorizing method bodies. When we replace Word2Vec+CNN with Sent2Vec, we can prepare the inconsistent method name detection system about 14 times faster than the conventional approach, without losing the detection performance in terms of precision, recall, and F value.

Our future work includes an enhancement of the inconsistent name detection performance. In this comparative study, we considered the alternatives only for the method-body vectorizing part. However, we can also vary the method-name vectorizing part approach. We plan to study the other available combinations of the

[3] https://deeplearning4j.org/.

vectorization approaches in the future. Furthermore, because various novel vectorization approaches have been developed in natural language processing studies [25], we also consider their applications to the method name evaluation as our significant future work.

Acknowledgements The authors would like to thank the anonymous reviewers for their helpful comments on an earlier version of this paper. This work was supported by JSPS KAKENHI Grant #20H04184, #21K11831, and #21K11833.

References

1. Liblit, B., Begel, A., Sweetser, E.: Cognitive perspectives on the role of naming in computer programs. In: Proceedings of 18th Annual Psychology of Programming Workshop, pp. 53–67 (2006)
2. Martin, R.C.: Clean Code: A Handbook of Agile Software Craftsmanship. Prentice Hall, Boston (2008)
3. Deissenboeck, F., Pizka, M.: Concise and consistent naming. Softw. Quality J. **14**(3), 261–282 (2006)
4. Boswell, D., Foucher, T.: The Art of Readable Code: Simple and Practical Techniques for Writing Better Code. Oreilly & Associates, Sebastopol, CA (2011)
5. Gosling, J., Joy, B., Steele Jr., G.L., Bracha, G., Buckley, A.: The Java Language Specification. Addison-Wesley, Boston, MA (2014)
6. Montgomery, S.L. (ed.): MISRA C: Guidelines for the Use of the C Language in Critical Systems 2012. Motor Industry Research Association, Warwickshire (2013)
7. Kernighan, B.W., Pike, R.: The Practice of Programming. Addison-Wesley Longman, Boston, MA (1999)
8. Free Software Foundation: GNU Coding Standards. https://www.gnu.org/prep/standards/. Accessed 20 Nov 2020
9. Allamanis, M., Barr, E.T., Bird, C., Sutton, C.: Learning natural coding conventions. In: Proceedings of 22nd ACM SIGSOFT International Symposium on Foundations of Software Engineering, pp. 281–293 (2014)
10. Liu, K., Kim, D., Bissyandé, T.F., Kim, T., Kim, K., Koyuncu, A., Kim, S., Traon, Y.L.: Learning to spot and refactor inconsistent method names. In: Proceedings of 41st International Conference on Software Engineering, pp. 1–12 (2019)
11. Mikolov, T., Sutskever, I., Chen, K., Corrado, G.S., Dean, J.: Distributed representations of words and phrases and their compositionality. Adv. Neural Inf. Process. Syst. **26**, 3111–3119 (2013)
12. Le, Q., Mikolov, T.: Distributed representations of sentences and documents. In: Proceedings of 31st International Conference on Machine Learning, vol. 32(2), pp. 1188–1196 (2014)
13. Goodfellow, I., Bengio, Y., Courville, A.: Deep Learning. MIT Press, MA (2016)
14. Liu, K.: Debug-Method-Name. https://github.com/SerVal-DTF/debug-method-name/tree/master/Data. Accessed 20 Dec 2020
15. Pagliardini, M., Gupta, P., Jaggi, M.: Unsupervised learning of sentence embeddings using compositional n-gram features. In: Proceedings of 2018 Conference on North American Chapter of the Association for Computational Linguistics: Human Language Technologies, vol. 1, pp. 528–540 (2018)
16. Høst, E.W., Østvold, B.M.: Debugging method names. In: Drossopoulou, S. (ed.) ECOOP 2009—Object-Oriented Programming. Lecture Notes in Computer Science, vol. 5653, pp. 294–317. Springer, Berlin, Heidelberg (2009)

17. Runeson, P., Alexandersson, M., Nyholm, O.: Detection of duplicate defect reports using natural language processing. In: Proceedings of 29th International Conference on Software Engineering, pp. 499–510 (2007)
18. Lawrie, D., Feild, H., Binkley, D.: Extracting meaning from abbreviated identifiers. In: Proceedings of 7th IEEE International Working Conference on Source Code Analysis & Manipulation, pp. 213–222 (2007)
19. Hill, E., Fry, Z.P., Boyd, H., Sridhara, G., Novikova, Y., Pollock, L., Vijay-Shanker, K.: AMAP: automatically mining abbreviation expansions in programs to enhance software maintenance tools. In: Proceedings of 2008 International Working Conference Mining Software Repositories, pp. 79–88 (2008)
20. Corazza, A., Martino, S.D. , Maggio, V.: Linsen: an efficient approach to split identifiers and expand abbreviations. In: Proceedings of 28th IEEE International Conference on Software Maintenance, pp. 233–242 (2012)
21. Alatawi, A., Xu, W., Yan, J.: The expansion of source code abbreviations using a language model. In: Proceedings of 2018 IEEE 42nd Annual Computer Software & Applications Conference, vol. 2, pp. 370–375 (2018)
22. Sauer, C., Jeffery, D.R., Land, L., Yetton, P.: The effectiveness of software development technical reviews: a behaviorally motivated program of research. IEEE Trans. Softw. Eng. **26**(1), 1–14 (2000)
23. Rigby, P., Cleary, B., Painchaud, F., Storey, M.-A., German, D.: Contemporary peer review in action: lessons from open source development. IEEE Softw. **29**(6), 56–61 (2012)
24. Bojanowski, P., Grave, E., Joulin, A., Mikolov, T.: Enriching word vectors with subword information. Trans. Assoc. Comput. Linguist. **5**, 135–146 (2017)
25. Devlin, J., Chang, M.-W. , Lee, K., Toutanova, K.: BERT: pre-training of deep bidirectional transformers for language understanding. In: Proceedings of 2019 Conference on North American Chapter of the Association for Computational Linguistics: Human Language Technologies, vol. 1, pp. 4171–4186. Association for Computational Linguistics, Minneapolis, Minnesota (2019)

Heart Sound Segmentation Based on a Joint HSMM Method

Yao Zhang, Zeyu Ma, Xin Zhou, Xianhong Li, Ying Liu, Mingang Chen, Xin Sun, Xuying Wang, Jingtao Wang, Lizhi Cai, and Kun Sun

Abstract In this study, we propose a novel Joint-HSMM method that combines the CNN and probabilistic models (HSMM) for heart sound segmentation. First, we build a semantic segmentation model and a time series probability model based on U-Net and HSMM respectively, and fusion model results by average consistency. The proposed method are conducted on various datasets, including the publicly available PhysioNet and a children heart sound dataset collected by Shanghai Xinhua Hospital. For the second dataset, we make use of Transfer Learning techniques for early convergence and robustness. It is shown that our method has outperformed the current state-of-the-art method by achieving the average recall, F1-score of 96.29%, 96.14% on PhysioNet dataset, and 86.95% and 87.24% on children heart sound dataset respectively.

Keywords Heart sound segmentation · Convolutional neural networks · Hidden semi-Markov models · PhysioNet · Model fusion

Y. Zhang · X. Li · X. Wang
Enterprise Research Institute, Hangzhou Ewell Technology Company, Hangzhou 310053, China

Z. Ma · M. Chen · L. Cai (✉)
Shanghai Key Laboratory of Computer Software Testing & Evaluating, Shanghai Development Center of Computer Software Technology, Shanghai 201112, China
e-mail: clz@sscenter.sh.cn

X. Zhou · X. Sun
Clinical Research Unit, Xinhua Hospital affiliated to Shanghai Jiaotong University School of Medicine, Shanghai 200092, China

Y. Liu · K. Sun (✉)
Department of Pediatric Cardiology, Xinhua Hospital affiliated to Shanghai Jiaotong University School of Medicine, Shanghai 200092, China
e-mail: drkunsun@xinhuamed.com.cn

J. Wang
Xunyin Intelligent Technology (Shanghai) Co., Ltd., Shanghai 200335, China

R. Lee (ed.), *Computer and Information Science 2021—Summer*,
Studies in Computational Intelligence 985,
https://doi.org/10.1007/978-3-030-79474-3_10

1 Introduction

The auscultation of heart sounds is still the main method of early screening for cardiovascular disease (CVD) [1]. Effective auscultation diagnosis often requires doctors to have professional medical knowledge and rich clinical auscultation experience, and it takes a long time to train a cardiovascular disease diagnosis expert, which leads to a serious shortage of cardiovascular disease experts, especially in non-Third-A hospitals. Extensive training is required to properly master cardiac auscultation and it was observed that, on average, only 20% of medical interns can effectively detect heart conditions using auscultation [2]. Therefore, the clinical need for effective intelligent aids based on heart sound signals (PCGs) is becoming more and more urgent.

The research on automatic identification of cardiovascular disease types based on heart sound signals has been started more than 50 years ago [3, 4], but it still faces many challenges. The Segmentation of the fundamental heart sound (FHSs) within a phonocardiogram (PCG), is an essential step in the automatic analysis of heart sounds. The accurate localization of the FHSs is prerequisite for the identification of the systolic or diastolic regions, allowing the subsequent classification of pathological situations in these regions [5]. Fundamental heart sounds (FHSs) usually include the first (S1) and second (S2) heart sounds. S1 occurs at the beginning of isovolumetric ventricular contraction, when the mitral and tricuspid valves close due to the rapid increase in pressure within the ventricles. S2 occurs at the beginning of diastole with the closure of the aortic and pulmonic valves (as shown in Fig. 1). While the FHSs are the most recognizable sounds of the heart cycle, the mechanical activity of the heart may also cause other audible sounds, such as the third heart sound (S3), the fourth heart sound (S4), systolic ejection click (EC), mid-systolic click (MC), diastolic sound or opening snap (OS), as well as heart murmurs caused by the turbulent, high-velocity flow of blood. Common noise sources include endogenous or ambient speech, motion artifacts and physiological sounds such as intestinal and breathing sounds. Other physiological sounds of interest, such as murmurs, clicks, splitting of the FHSs and additional S3 and S4 sounds, can also complicate the identification of the FHSs [6]. For children, the segmentation and recognition of heart sound signals is more complicated: Firstly, children often cry during the process of collecting heart sound; secondly, parents' comforting sounds also hinders the clarity of heart sounds, furthermore, children's heart sounds are weaker than adults' heart sounds. For better segmentation of PCG signal FHSs, we propose to combine the semantic segmentation model based on convolutional neural network and the sequence probability model LR-HSMM [6] to construct a joint heart sound segmentation model. Firstly, we preprocess the PCG signal, and extract the four signal envelope features of homomorphic envelogram, Hilbert envelope, wavelet envelope and power spectral density (PSD) envelope. Then the semantic segmentation model and sequence probability model are trained separately. Finally, a joint segmentation model is constructed by weighted joint. We first train and evaluate the model on the current largest open source heart sound data set, the PhysioNet 2016 dataset, and then apply the transfer learning solution to complete the FHSs segmentation modeling and evaluation

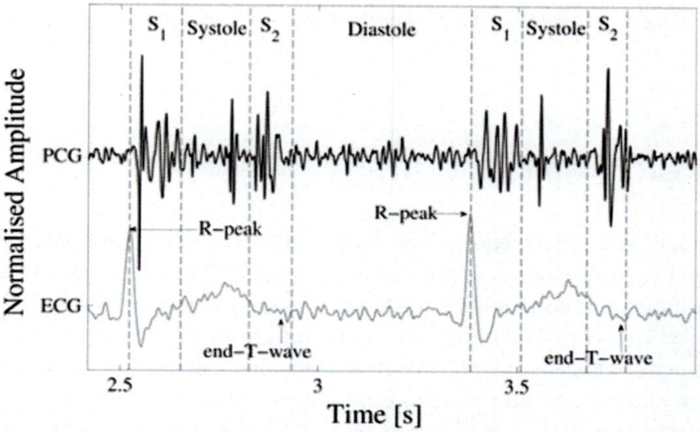

Fig. 1 Example of an ECG-labelled PCG, with the ECG, PCG and four states of the heart cycle (S1, systole, S2 and diastole) shown. The R-peak and end-T-wave are labeled as references for defining the approximate positions of S1 and S2 respectively. Mid-systolic clicks, typical of mitral valve prolapse, can be seen

of children's heart sound PCG signals. The joint algorithm we proposed obtained the state-of-the-art effect on the PhysioNet2016 dataset, and also achieved a good segmentation effect on the complex children's heart sound data.

In this paper, related works are surveyed in Sect. 2; the proposed segmentation algorithm flow is presented in Sect. 3. In Sect. 4, we report our experimental methodology and results on the aforementioned datasets this section also comes to the conclusions and puts forward future works.

2 Related Work

2.1 *Heart Sound Segmentation*

Many researchers extract the time domain, frequency domain and envelope characteristics of the PCG signal, and then use the amplitude threshold to perform segmentation detection of the first heart sound and the second heart sound [7–11]. This kind of method requires the researcher's high feature extraction ability. In addition, the hard threshold method is adopted, and the generalization ability of the obtained model is poor, especially for abnormal samples. Then, various types of classifiers have been proposed in the literature. From simple k-means clustering [12, 13] to decision trees [14], from support vector machine (SVM) classifiers [11], to artificial neural network(ANN) classifiers [15, 16]. More recently, Chen et al. [17] proposed the use of fully connected deep neural networks to distinguish S1 sounds from S2 sounds.

2.2 Model Training

The introduction of the sequence probability model greatly improves the segmentation effect of heart sounds. Gamero and Watrous [18] extracted mel-frequency cepstral coefficients features and using hidden Markov models (HMMs). Ricke et al. [19] used embedded HMMs within each state, along with mel-frequency cepstral coefficients, Shannon energy and regression coefficients. Gill et al. [20] proposed combine HMMs with heart sound FHS timing durations, which yielded high segmentation accuracy on a small dataset. Schmidt et al. [21] were the first researchers to explicitly model the expected duration of heart sounds within the HMM using a HSMM. David B. Springer extension of the Viterbi algorithm and more discriminative emission probability derivation using logistic regression(LR) within the HSMM, this method to be the current state-of-the-art model.

This work builds a joint-HSMM algorithm which combined CNN with sequence probability model. Our method outperform current state-of-the-art heart sound segmentation method in PhysioNet2016 dataset and it has achieved good results in the segmentation of children's heart sound signals.

3 Methodology

In this section, we firstly described the datasets used in our experiment, then the four main steps of the proposed segmentation algorithms are described in detail which includes data preprocessing, model training and model fusion mechanism, at last, we explain the quantitative indicators for evaluating experimental results.

3.1 Data Preprocessing

The same preprocessing process described in [6] is applied to the original PCG signals are downsampled 1000 Hz, filtered with high-pass and low-pass Butterworth filters, and the spike removal procedure. Then the four envelograms/envelopes features considered in [6] are extracted from the preprocessed signals, specifically homomorphic envelogram, Hilbert envelope, wavelet envelope and power spectral density (PSD) envelope. Finally, all the four envelograms/envelopes are downsampled 50 Hz and then normalized. The normalized envelograms are used as the feature representation of the original heart sound signal and are collected in the four-dimensional signal, where for $t = 0, \ldots, T - 1$, and where t denotes the time instant 50 Hz. Then, $s(t)$

is defined as the state labels sequence associated to each time instant, i.e., where state 0 corresponds to an S1 sound, state 1 to a systole interval or a diastole interval, and state 2 to an S2 sound.

3.2 Model Training

Cardiac cycle segmentation is naturally related to image semantic segmentation in computer vision field in which image semantic segmentation can be considered to classify each pixel in an image, while cardiac cycle segmentation can be thought of as a semantic segmentation task of a continuous one-dimensional heart sound signal by determining whether each sample point of signal belongs to S1, systolic, S2, or diastolic phases. In addition, the U-Net algorithm is widely used in image semantic segmentation, especially in medical image semantic segmentation, so we adopt the CNN structure in a variant based on U-Net [22] as one of the models of cardiac cycle segmentation.

Compared with the original U-Net, this variant CNN structure modifies the two-dimensional convolution in the original architecture into an one-dimensional convolution to adapt to the one-dimensional heart sound signal; and reduces the number of channels in each convolutional layer, which increases the computational speed of the algorithm and reduces the over-fitting risk of model to a certain extent as a result of the decrease of the model complexity. Furthermore, in this work, the systolic and diastolic periods are grouped into one category, so the number of categories of CNN output layer is 3 instead of 4 in [22]. It should be noted that the U-Net mentioned below refers to the modified U-Net variant unless otherwise specified.

In order to balance the computational complexity of CNN model and the input sequence length containing at least 3 cardiac cycles, so that there is enough information for cardiac cycle segmentation, the final feature length was selected as 128, which is equivalent to 2.56 s of heart sound signal. We split the original feature sequence into a series of feature patches in a sliding window way with a window size of 128 without overlapping and the last window with a size less than 128 is indented forward to obtain a 128 full size window. According to the same clipping method, the state label sequence corresponding to each feature window is obtained as the ground truth labels of U-Net semantic segmentation. Therefore, the actual input data into the proposed CNN network is a series of 4×128 feature patches, and the corresponding segmentation label is a sequence of state labels with a length of 128, i.e. $s(t) \in \{0, 1, 2\}$. Our model structure has been show on Fig. 2.

The model training of U-Net is performed in two stages: pre-training and transfer learning. In the first stage, the PhysioNet/CinC challenge 2016 data set is divided into fivefold cross validation data sets on the patient basis, and the model training and performance evaluation are performed to obtain the best model hyperparameters. Then, all the data of Physio-Net 2016 are used for the training the model, and the model with the best performance metrics on the test set of private hospital data as the pre-trained model of the second stage. In the second stage, the parameters of the

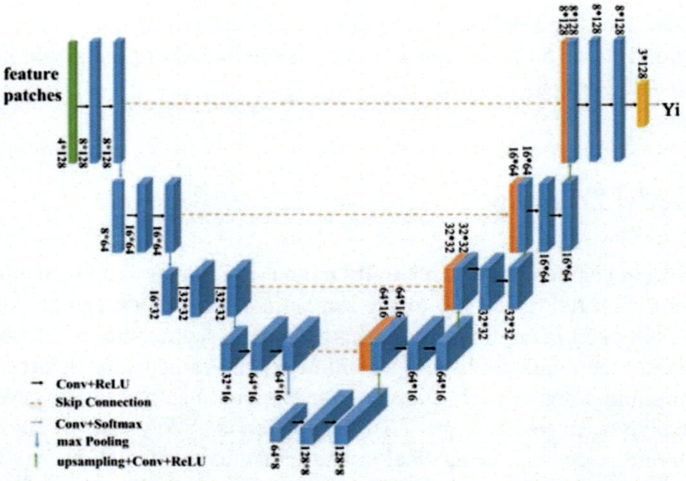

Fig. 2 Schematic representation of the U-Net structure, here we take batch size to be 1. In the last rectangular block as in the yellow part is a 3-classes softmax, which classifying S1 hear sound, S2 heart sound and the cluster of systolic and diastolic periods, the number 4 at the beginning of the model is the amount of feature envelope maps

pre-training model obtained in the first stage are used as the initialization parameters, and the training set of XinHua data is used to fine-tune the model to better fit the data set. The model with the best performance on the XinHua test set at this stage is selected as the model for subsequent algorithm comparison and model fusion.

The current state-of-the-art algorithm LR-HSMM [6] in the field of cardiac cycle segmentation is selected as another base model for the fusion model, and the data preprocessing and model training of LR-HSMM are operated as in the original paper. In particular, it should be pointed out that the modified U-Net variant and LR-HSMM in this paper use the same data preprocessing and feature extraction process, only the subsequent state sequence prediction method is different. The U-Net variant directly uses an one-dimensional deep convolutional neural network to implement semantic segmentation from feature patches to state sequences, while LR-HSMM uses LR for state classification and HSMM-based temporal modeling as described in [6] based on feature patches.

3.3 Model Fusion Mechanism

For the joint sequence probability models ability to mine timing information and CNN's ability to extract local context information for modeling, this study proposes a different model fusion mechanism from Renna et al. [22], And achieved state-of-the-art effects on the task of heart sound segmentation. Francesco Renna simply replaces logistic regression(LR) with CNN to obtain the emission probability of the sequence probability model. From our experimental research, it is found that the key point is not here. Simple logistic regression is enough to get a good emission probability. To make full use of the characteristics of the CNN model and the sequence probability model, the results should be weighted and fused at the end of the Viterbi decoding algorithm, so that the complementarity of the two algorithms can be better utilized to improve the final segmentation effect. The entire flow of the algorithm is shown in Fig. 3.

As shown in Fig. 3, the PCG signal is preprocessed to extract four envelope features, and the CNN model and the LR-HSMM sequence probability model are trained respectively. In the U-Net modeling branch, first pass the State Encode process to perform the four-state encoding of S1, systole, S2, and diastole on the output of U-Net to obtain the initial sequence of state encoding Seq1. In the BiDecode process, Seq1 is decoded in both directions so that the state sequence meets the four-state sequential connection of S1, systole, S2, and diastole. In this process, Forward-Decode state sequence Seq2 and Backward-Decode state sequence Seq3 are obtained respectively.

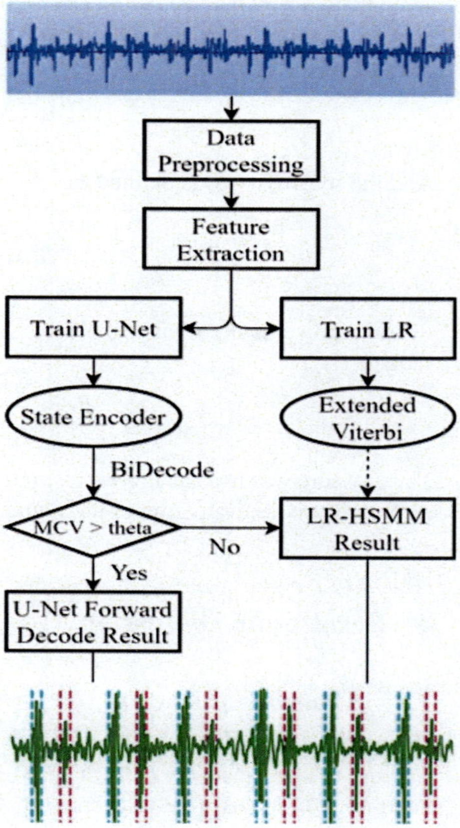

Fig. 3 Schematic diagram of model fusion, mean consistency value (MCV), θ is 0.98

Then calculate the mean consistency value (MCV), When the MCV is greater than 0.98, the Unet-Forward-Decode State Sequence is used as the final segmentation prediction result. Otherwise, the State Sequence decoded by LR-HSMM using the Expand-Viterbi algorithm is used as the final prediction result. In particular, on denoting by y_U the output of U-Net, by y_{IH} the output of LR-HSMM, by I the Schematic function, our final result can be expressed as:

$$y = y_U \cdot I_{MCV>\theta} + y_{IH} \cdot (1 - I_{MCV>\theta}) \tag{1}$$

3.4 Performance Metrics

The performance of the segmentation algorithms were evaluated using positive predictive value (P_+), sensitivity (S_e) and score, which is defined as follows. First, a true positive (T_p) is counted when the center of an S1(S2) sound in the estimated sequence is closer than 100 ms from the center of the corresponding S1(S2) sound in the ground truth sequence. All other S1 and S2 sounds in the estimated state sequence are considered as false positives (F_p). Then, the positive predictive value is defined as:

$$P_+ = \frac{T_p}{T_p + F_p} \tag{2}$$

And the sensitivity S_e is defined as:

$$S_e = \frac{T_p}{T_{total}} \tag{3}$$

where represents the total number of S1 and S2 sounds in the ground truth state sequence $s(t)$.

$$F_1 = \frac{2 \times P_+ \times S_e}{P_+ + S_e} \tag{4}$$

The F_1 score was used as it gives a single harmonic mean of P_+ and S_e, which gives a comprehensive evaluation of the segmentation results.

4 Results and Discussion

4.1 Experiment Result

In order to investigate the performance of the proposed CNN-based algorithms over heart sound signals, five-fold cross-validation experiments are performed on the

Table 1 The segmentation performance comparison of five-fold cross-validation on PhysioNet2016 dataset in percentage (%)

Methods	P_+	S_e	F_1
U-Net[1]	95.89±1.64	95.04±1.79	95.46±1.70
LR-HSMM[2]	93.68±1.25	95.27±1.23	94.46±1.22
Joint Model	**95.98± 1.04**	**96.29 ± 0.91**	**96.14 ± 0.97**

Table 2 The segmentation performance comparison on hospital private dataset in percentage (%)

Methods	P_+	S_e	F_1
U-Net[1]	86.64	85.83	86.23
Fine-tuned U-Net[2]	**87.77**	86.92	**87.34**
LR-HSMM[3]	84.62	81.61	83.09
Joint Model[1+3]	86.15	85.61	85.88
Joint Model[2+3]	87.54	**86.95**	87.24

publicly available dataset PysioNet2016. Table 1 reports the mean and the standard deviation of the performance metric values corresponding to the different segmentation algorithms considered in this work. The proposed algorithm achieves an average score of 95.46 ± 1.70%, of 95.89 ± 1.64% and of 95.04 ± 1.79% on the validation datasets, which outperforms all the evaluation metrics of the state-of-the-art algorithm LR-HSMM (94.46 ± 1.22%, 93.68 ± 1.25% and 95.27 ± 1.23%, respectively). Furthermore, the joint model integrated U-Net and LR-HSMM achieves the best metrics (96.14 ± 0.97%, 95.98 ± 1.04% and 96.29 ± 0.91%, respectively) using optimal consistency threshold 0.98 obtained from the cross-validation experiments, which indicates the fusion method takes advantage of the complementarity of the above two algorithms to a certain extent (Table 2).

4.2 Discussion

The introduction of sequence probability models, such as HMM, HSMM [19–21], especially the LR-HSMM [6] algorithm proposed by David B. Springer to add heart sound FHSs duration information in the Viterbi process, which occupied the state-of-the-art position for a long time. Through our experimental analysis, it is found that the sequence probability model of LR-HSMM has two obvious disadvantages:

1. The algorithm performance reaches the bottleneck earlier. It is difficult to improve the model effect by simply increasing the complexity of the data. The key lies in the extracted features, and it is difficult to improve the model performance.
2. This type of model is prone to predict errors in the entire sequence due to a prediction error of one state of the preamble.

Based on the above findings, we propose the idea of constructing a model by combining CNN and LR-HSMM. Renna et al. [22] also proposed the idea of joint CNN and sequence probability model to do heart sound FHSs segmentation, but compared to LR-HSMM, they only replace LR with CNN to build a series of fusion models such as CNN-HMM and CNN-HSMM. Judging from the results, this fusion method does not have the effect of learning from each other. As far as PCG signal segmentation is concerned, it is necessary to introduce the characteristics of the sequence probability model algorithm. The key reason why the fusion method proposed by Renna et al. [22] cannot greatly improve the effect of the model is that it is sufficient to use logistic regression to calculate the emission probability when the feature engineering is basically unchanged.

The fusion method proposed in this study significantly improves the segmentation effect of heart sound signals. It has certain reference value for the subsequent improvement of the effect of heart sound segmentation and the recognition of abnormal heart sounds. However, this research still has certain limitations. First of all, the sample size used is not large enough, and the generalization ability of the model is limited. Furthermore, this research still depends on the quality of features. Good features are still critical to the entire model system. Thus how to extract better features from PCG signals is still the focus of follow-up research.

5 Conclusion

In this paper, we propose an novel heart sound segmentation algorithm based on model fusion, which uses CNN and LR-HSMM for separate training and MCV for the fusion of results. The results show that this method can perform effective heart sound segmentation and outperformed the current state-of-the-art method. At the application level, we tested our method the open source data set PhysioNet and the Xinhua children's heart sound dataset. In addition, we used transfer learning techniques in the experiments on the latter dataset, which shows the robustness of our method.

The possible improvement options based on the proposed method are mainly about: firstly, finding a better model fusion strategy to improve the effect; secondly, performing a deeper fusion within the model in order to perform a complementation of the CNN and LR-HSMM.

Acknowledgements This work is funded by Investigator Initiated Trial (IIT) Program (No. 19XHCR140) Exploration and application of artificial intelligence technology-assisted heart sound recognition system for congenital heart disease screening in children.

References

1. Reed, T.R., Reed, N.E., Fritzson, P.: Heart sound analysis for symptom detection and computer-aided diagnosis. Simul. Model. Pract Theory **12**(2), 129–146 (2004)
2. Mangione, S.: Cardiac auscultatory skills of physicians-in-training: a comparison of three English-speaking countries. Am. J. Med. **110**(3), 210–216 (2001)
3. Gerbarg, D.S., Taranta, A., Spagnuolo, M., Hofler, J.J.: Computer analysis of phonocardiograms. Prog. Cardiovasc. Dis. **5**(4), 393–405 (1963)
4. Liang, H., Nartimo, I.: A feature extraction algorithm based on wavelet packet decomposition for heart sound signals. In: Proceedings of the IEEE-SP International Symposium on Time-Frequency and Time-Scale Analysis (Cat. No. 98TH8380), pp. 93–96. IEEE (1998)
5. Leatham, A.: Auscultation and phonocardiography: a personal view of the past 40 years. Br. Heart J. **57**(5), 397 (1987)
6. Springer, D.B., Tarassenko, L., Clifford, G.D.: Logistic regression-HSMM-based heart sound segmentation. IEEE Trans. Biomed. Eng. **63**(4), 822–832 (2015)
7. Liang, H., Lukkarinen, S., Hartimo, I.: Heart sound segmentation algorithm based on heart sound envelogram. In: Computers in Cardiology, pp. 105–108. IEEE (1997)
8. Kumar, D., Carvalho, P., Antunes, M., Henriques, J., Eugenio, L., Schmidt, R., Habetha, J.: Detection of S1 and S2 heart sounds by high frequency signatures. In: 2006 International Conference of the IEEE Engineering in Medicine and Biology Society, pp. 1410–1416. IEEE (2006)
9. Ari, S., Kumar, P., Saha, G.: A robust heart sound segmentation algorithm for commonly occurring heart valve diseases. J. Med. Eng. Technol. **32**(6), 456–465 (2008)
10. Vepa, J., Tolay, P., Jain, A.: Segmentation of heart sounds using simplicity features and timing information. In: 2008 IEEE International Conference on Acoustics, Speech and Signal Processing, pp. 469–472. IEEE (2008)
11. Vepa, J.: Classification of heart murmurs using cepstral features and support vector machines. In: 2009 Annual International Conference of the IEEE Engineering in Medicine and Biology Society, pp. 2539–2542. IEEE (2009)
12. Gupta, C.N., Palaniappan, R., Swaminathan, S., Krishnan, S.M.: Neural network classification of homomorphic segmented heart sounds. Appl. Soft Comput. **7**(1), 286–297 (2007)
13. Chen, T., Kuan, K., Celi, L.A., Clifford, G.D.: Intelligent heartsound diagnostics on a cellphone using a hands-free kit. In: 2010 AAAI Spring Symposium Series (2010)
14. Stasis, A.C., Loukis, E., Pavlopoulos, S., Koutsouris, D.: Using decision tree algorithms as a basis for a heart sound diagnosis decision support system. In: 4th International IEEE EMBS Special Topic Conference on Information Technology Applications in Biomedicine, pp. 354–357. IEEE (2003)
15. Hebden, J.E., Torry, J.: Neural network and conventional classifiers to distinguish between first and second heart sounds (1996)
16. Sepehri, A.A., Gharehbaghi, A., Dutoit, T., Kocharian, A., Kiani, A.: A novel method for pediatric heart sound segmentation without using the ECG. Comput. Methods Programs Biomed. **99**(1), 43–48 (2010)
17. Chen, T.E., Yang, S.I., Ho, L.T., Tsai, K.H., Chen, Y.H., Chang, Y.F., Lai, Y.H., Wang, S.S., Tsao, Y., Wu, C.C.: S1 and S2 heart sound recognition using deep neural networks. IEEE Trans. Biomed. Eng. **64**(2), 372–380 (2016)
18. Gamero, L., Watrous, R.: Detection of the first and second heart sound using probabilistic models. In: Proceedings of the 25th Annual International Conference of the IEEE Engineering in Medicine and Biology Society (IEEE Cat. No. 03CH37439), vol. 3, pp. 2877–2880. IEEE (2003)
19. Ricke, A.D., Povinelli, R.J., Johnson, M.T.: Automatic segmentation of heart sound signals using hidden Markov models. In: Computers in Cardiology, pp. 953–956. IEEE (2005)
20. Gill, D., Gavrieli, N., Intrator, N.: Detection and identification of heart sounds using homomorphic envelogram and self-organizing probabilistic model. In: Computers in Cardiology, pp. 957–960. IEEE (2005)

21. Schmidt, S.E., Holst-Hansen, C., Graff, C., Toft, E., Struijk, J.J.: Segmentation of heart sound recordings by a duration-dependent hidden Markov model. Physiol. Measur. **31**(4), 513 (2010)
22. Renna, F., Oliveira, J., Coimbra, M.T.: Deep convolutional neural networks for heart sound segmentation. IEEE J. Biomed. Health Inf. **23**(6), 2435–2445 (2019)

A Novel Authentication System for Artwork Based on Blockchain

Seung Wook Jung

Abstract The blockchain technology is emerging technology and various applications adapt blockchain. Also, art domain tries to adapt the blockchain technology because the integrity of artwork information and the authenticity of artwork are crucial factors for the price of artwork. The blockchain supports these properties well. However, existing services based on blockchain in art domain could not address the problem so called, blockchain air gap, between the blockchain listing and physical artworks. This paper provides the novel solution that addresses the problem efficiently but not expensively. Additionally, proposing system does not store owner's password, which is used for proving the ownership and the authenticity of artwork, in system at all. Therefore, proposing system is very secured.

Keywords Blockchain · Authenticity · Ownership · Art

1 Introduction

Blockchain is a distributed ledger technology (DLT) which provides the integrity and the irreversibility of the records stored in the distributed ledger. Therefore, many applications of various domains including art domain are emerging. In the art domain, the authenticity and the provenance of artworks are vital factors to decide the price of the artworks, so the serval applications try to adopt blockchain to the art domain [1–6].

There are three ways to authenticate the artworks: (1) personal expertise, (2) scientific analysis, and (3) the certificate of authenticity. This paper focuses on the certificate of authenticity [7].

Some applications based on blockchain focus on galleries or museums as the customers. The others are the artists. This paper focuses on the artists as the customers.

S. W. Jung (✉)
Department of Cyber Security, Konyang University, Nonsan, Korea
e-mail: swjung@konyang.ac.kr

R. Lee (ed.), *Computer and Information Science 2021—Summer*,
Studies in Computational Intelligence 985,
https://doi.org/10.1007/978-3-030-79474-3_11

This paper proposes a novel method to authenticate artworks and to prove the ownership. The blockchain is very attractive system for storing the ownership information and the information of the authenticity of artwork by the artist because the integrity of records is guaranteed, and the first-mile problem is protected. The first-mile problem is that the garbage inputs, the garbage is protected by the blockchain.

This paper proposes a novel blockchain-based the authenticity of artworks and the proving ownership. In this paper, an artist, who is verified by Know Your Customer (KYC), has a private key and a public key. The public key is stored in the proposed system, probably in blockchain. The artist stores the information of artwork with the signature using private key. After that, the artist puts the password for an artwork, and then the proposing system generates QR code for the artwork. The QR code will be attached or printed at the certificate of authenticity. When the artist sales and hands over the artwork, the certificate of authenticity is also handed over to a buyer. The buyer will put his own password and the system will change the password related with the QR code. After that, the owner will prove the ownership and the authenticity with the QR code and the password.

The QR code is unchangeable but the password can be changeable by the owner. The password is not stored in the online system. Therefore, the system is secured. The paper is the first attempt to address the challenge of managing the "blockchain air gap", between the blockchain listing and the physical artwork.

This paper is organized as following. In Sect. 2, we review the previous works in blockchain for art domain. Section 3 describes our system and Sect. 4 analyzes the security of our system. In Sect. 5 we discuss on our contribution, and finally conclude in Sect. 6.

2 Related Works

In artwork domain, following areas have used blockchain for solving each problem of traditional artwork market [7].

1. Blockchain is used as provenance and authenticity registries.
2. Digital scarcity [8].
3. Fractional equity and shared ownership [4].
4. New forms of copyright registry [6].

Provenance and authenticity are closely related in blockchain because provenance describes the chain of ownership. The ownership is proved by authentication. There are three methods of the authenticity of artwork: personal expertise, scientific analysis, and certificate of authenticity [7]. In this paper, we focus on the method of certificate of authenticity. That is, authentication is done by a certificate of authenticity. Without certificate, one cannot prove the ownership of the artwork.

There are serval blockchain-based art authentication services: Verisart, Artory and Codex Protocol.

Verisart launched in 2015 to provide a trusted database for artworks. However, this service is publicly accessible, so anyone can register artwork. For example, in June 2018, a man named Terence Eden listed the Mona Lisa on Verisart blockchain by himself as the artist and 1506 as the creation date. This problem is called first-mile problem [9]. This paper solves this problem using KYC in a way that the only KYCed artist can list artworks and if an artist lists false information, then our system is able to track who putted false information.

Artory only lists artworks that have been vetted by a partner organization such as auction house or a gallery, so Artory solves the problem of Verisart.

Codex Protocol also only lists artworks in blockchain vetted by partner.

However, Verisart, Artory, and Codex all have the challenge of managing "blockchain air gap", between the blockchain listing and the physical artwork [10]. This challenge has not been resolved yet [7]. This paper addresses this challenge.

Digital scarcity is done in a way that limited editions of digital art to track using blockchain [8].

When an artist sales his/her artwork at the first sale, the price is low. However, in the secondary market, the artwork can be awfully expensive. Therefore, [4] presents that the artist retains fractional equity in their work at the point of the first sale. This method uses the tokenization of artwork using blockchain technology. This paper can also support fractional equity even if it does not use tokenization technology.

Shared ownership of collaborative work is supported by blockchain and it is the new forms of copyright.

3 Proposed System

3.1 Overall System Architecture

Figure 1 shows the overall system architecture. Overall system consists of the artist, the owner of artwork, blockchain system, which is developed based on Hyperledger fabric, an off-chain web server, and a key storage.

First the artist sends artwork information and a password to the off-chain web server. The off-chain server generates a serial number and an encryption key (Key_0) for the artwork. The serial number is encrypted with Key_0. The password is hashed with SHA-256 hashing algorithm. The hashed value (Key_1) is XORed with Key_0 and the XORed result is Key_2. The off-chain server stores Key_2 in off-chain server and Key_0 in Key storage, which has no Internet connection, then the password is deleted in the memory in the off-chain server. The key storage has only one connection to the off-chain server with a specific TCP port and all other TCP ports are closed. Therefore, an attacker cannot access through Internet. The encrypted serial number is the content of QR code. The encrypted serial number, serial number and owner ID (here the artist ID) are stored in the blockchain. When the artist generates the certificate of authenticity, the QR code is printed in the certificate.

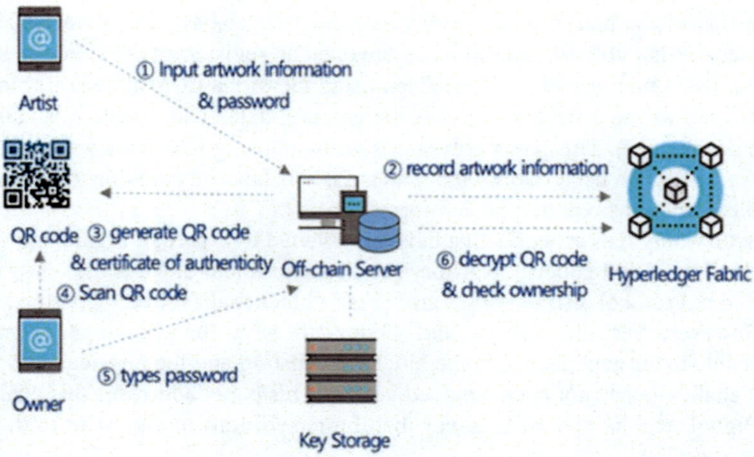

Fig. 1 Overall system architecture

If owner would want to verify the ownership and the authenticity of the artwork, the owner will scan QR code and type the password. The owner ID, QR code and typed password are sent to the off-chain server. The off-chain server generates Key_1 hashing the typed password and retrieves Key_2 with owner ID. The off-chain server sends the owner ID, Key_0 ($= Key\_1 \oplus Key\_2$) and encrypted serial number, i.e., QR code, to the blockchain system. The blockchain smart contract decrypts encrypted serial number and matches the serial number in the ledger with decrypted QR code. If matched and owner ID is right, then the blockchain returns verification success message to the off-chain server. The off-chain server passes the message to the owner.

3.2 Database, Key Storage and Blockchain Records

Figure 2 shows the database tables. The artwork table includes serial number, title, size, publishing year, artist name, materials, ownerID and encrypted Key_2. Key_2 is encrypted with Password-Based Encryption (PBE) using AES-256 algorithm based on administrator's password. OwnerID will be changed whenever owner is changed. User table is the same as ordinary user table. Figure 3 shows the table in the key storage. Key_0 is encrypted with PBE with the administrator's password. Figure 4 shows the blockchain record including encrypted serial number, artwork serial number and ownerID list.

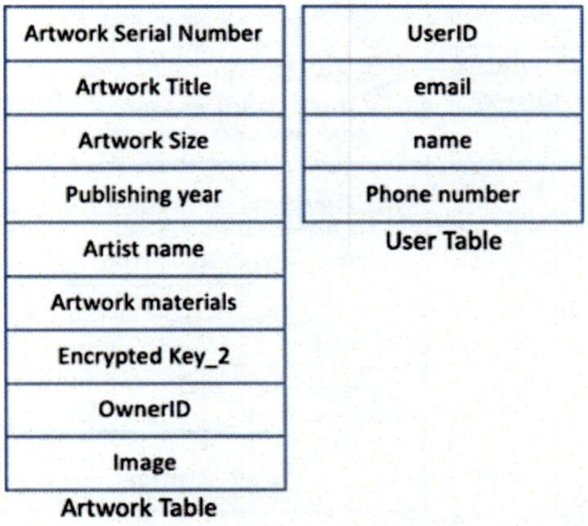

Fig. 2 Tables in the off-chain server

Fig. 3 Tables in key storage

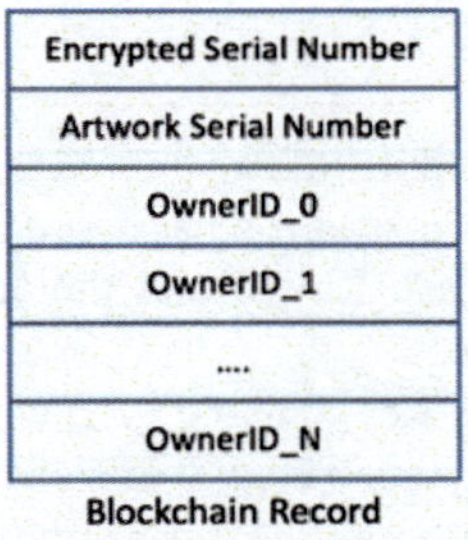

Fig. 4 Blockchain record

3.3 Issuing QR Code Process

The issuing QR code process is as following and shown in Fig. 5.

1. The artist takes pictures, and types title, size, publishing year, artist name, materials and passwords.
2. The off-chain server stores the information with serial number, which is randomly chosen, in database. Also, the off-chain server stores the artist's userID as ownerID.
3. The off-chain server generates a random key Key_0 and hashing the password (Key_1).

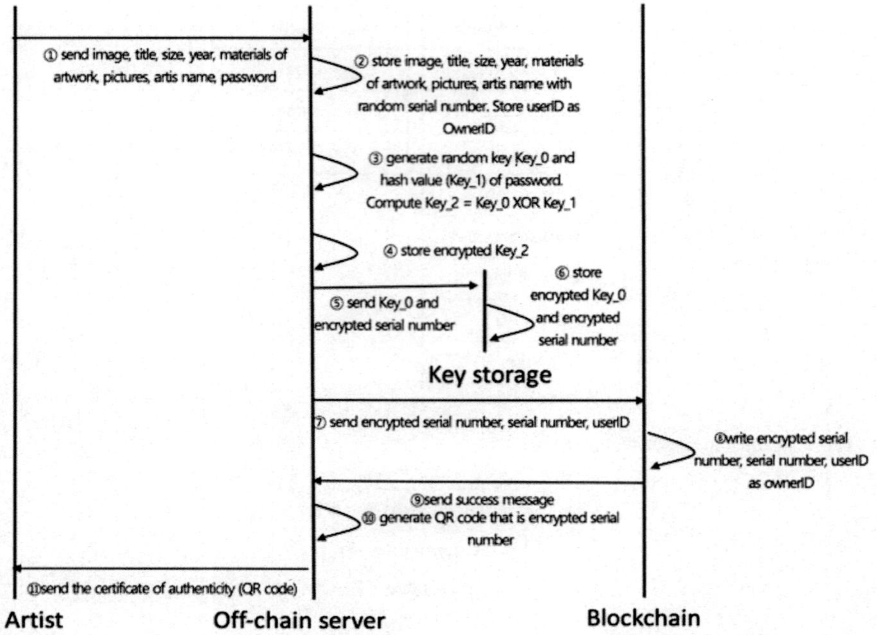

Fig. 5 Issuing QR code process

4. The off-chain server computes $Key\_0 \oplus Key\_1 = Key\_2$. Key_2 is encrypted using PBE with the administrator's password and stored in database.
5. The off-chain server sends Key_0 and the serial number encrypted with the Key_0 to the key storage.
6. The key storage encrypts Key_0 using PBE with the key storage administrator's password and stores encrypted Key_0 and encrypted serial number.
7. The off-chain server encrypts the serial number and sends blockchain smart contract the encrypted serial number, serial number and ownerID.
8. The blockchain smart contract writes encrypted serial number, serial number and ownerID on blockchain. Then, the blockchain system sends success message to the off-chain server.
9. The off-chain server generates QR code with the encrypted serial number and sends the certificate of authenticity including QR code to the artist.

The artist should be verified with KYC. For instance, the artist uploads a picture that includes his/her face and his/her passport. Also, the artist uploads his/her passport to the off-chain system. Then administrator checks whether (s)he is claimed artist. With KYC, the artist is authenticated. When the artist registers an artwork with false information, e.g., Mona Lisa with his/her name, the system can track it.

3.4 Verifying Ownership and Authenticity Process

The verification of ownership and the authenticity of the artwork are following and shown in Fig. 6.

1. Owner scans QR code in the certificate of authenticity and types the password.
2. The owner sends the off-chain server userID, encrypted serial number and the password.
3. The off-chain server retrieves encrypted Key_2 with userID and decrypts encrypted Key_2 with the administrator's password. The off-chain server computes Key_1 that is hash value of the password.
4. The off-chain server computes $Key\_0 = Key\_2 \oplus Key\_1$.
5. The off-chain server sends userID, Key_0 and encrypted serial number to the blockchain smart contract.
6. The blockchain smart contract decrypts encrypted serial number with Key_0
7. The blockchain smart contract retrieves the serial number and ownerID with encrypted serial number.
8. The blockchain smart contract checks whether the retrieved serial number is matched with decrypted value. Also, the blockchain smart contract checks whether the userID and ownerID are matched. If so, the blockchain smart contract sends verification success message to the off-chain server.
9. The off-chain server sends the result of verification to the owner.

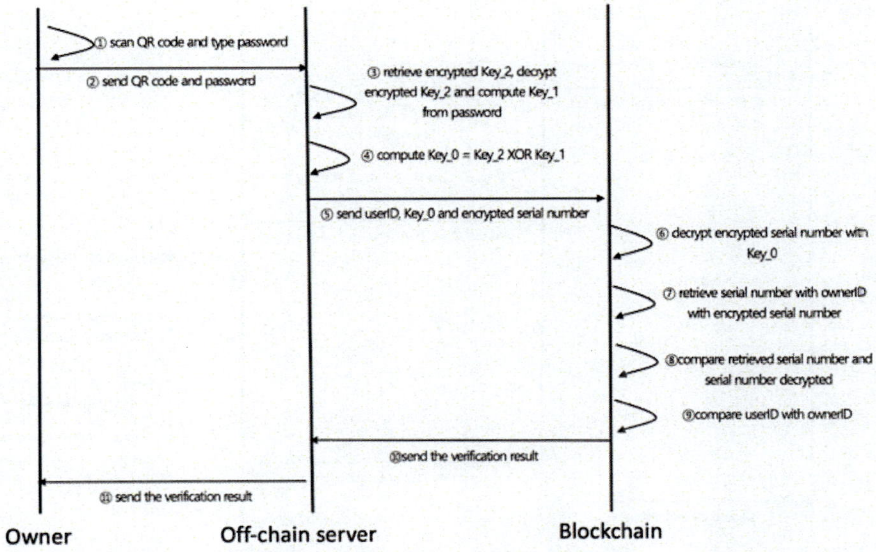

Fig. 6 Verifying ownership and authenticity process

The blockchain is trustworthy database, so our system stores the serial number, ownerID and encrypted serial number. The only who knows the right password can be verified as owner. Therefore, assume that a thief steals an artwork and the certificate of authenticity. However, the thief cannot be verified as owner because he does not know the password. Therefore, stealing is suppressed.

3.5 *Transferring Ownership Process*

The transferring ownership process is following and shown Fig. 7.

1. The verifying ownership and authenticity process is done. In sum, Key_0 is known by the off-chain server in this session.
2. If the ownership is verified, then the new owner types new password to the application in his/her smart phone
3. The application generates QR code including new password and new owner's ID.
4. The owner scans the QR code and sends the off-chain server it.
5. The off-chain server computes the hash value (Key_1) of new password.
6. The off-chain server computes $Key\_2 = Key\_0 \oplus Key\_1$ and encrypts Key_2 using PBE. Note Key_0 is computed at the verifying ownership and authenticity process.

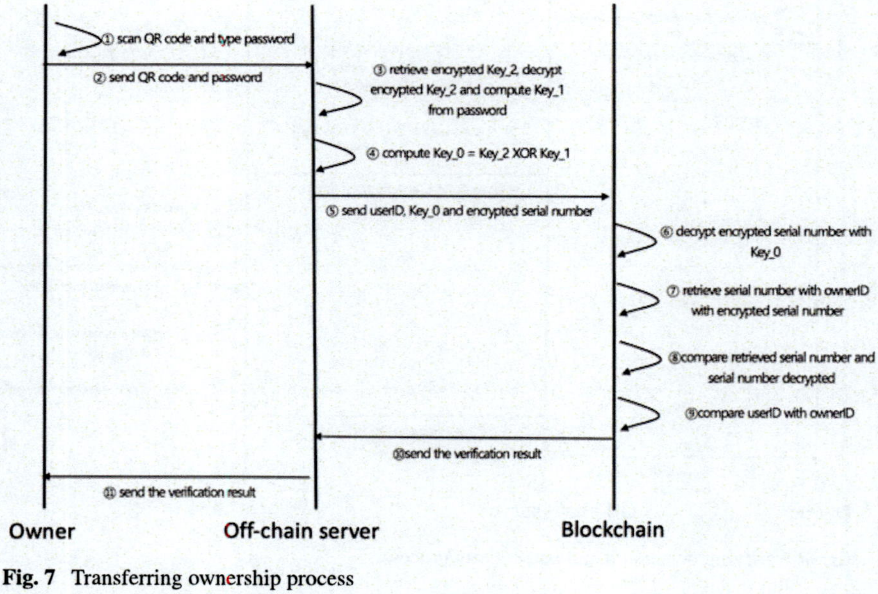

Fig. 7 Transferring ownership process

7. The encrypted Key_2 is stored in off-chain server's database and the off-chain server stores new userID as ownerID.
8. The off-chain server sends encrypted serial number and new owner's ID to the blockchain smart contract.
9. The smart contract writes new owner ID in the ledger.

If $Key\_0 = Key\_1 \oplus Key\_2$, then $Key\_2 = Key\_0 \oplus Key\_1$. Therefore, if $Key\_2 = Key\_0 \oplus Key\_1$, then $Key\_0 = Key\_1 \oplus Key\_2$. So that, transferring ownership works. In this way, the QR code printed in the certificate of authenticity does not need to be changed. However, the password can change.

Assume that owner has a genuine artwork, a forgery artwork, and the genuine certificate of authenticity. If (s)he sales the forgery artwork and the certificate of authenticity, at the sale point, new owner types his/her own password and the off-chain server changes Key_2. This means that the old owner cannot be verified ownership, because Key_2 is changed and he does not know new password, even if he has genuine artwork and the copy of certificate of authenticity. Forever he cannot claim that he has genuine artwork. Moreover, the chain of ownership is recorded in blockchain ledger, so (s)he can be tracked, if one knows that the artwork is the forgery. Therefore, doing so is suppressed.

Our system is the first novel solution that bridges "the blockchain air gap", between the blockchain listing and the physical artwork.

4 Security Analysis

The password is not stored in our system. This is different from ordinary web services that store hash value of password. If the attacker compromises the ordinary web services and gets the list of the hash value of passwords, the attacker tries to dictionary attack or brute force attack on the list of hash value. The attacker finds some of passwords and does not find the others.

In our system, the attacker must compromise the off-chain server first because the attacker can only access the off-chain server through Internet. The key storage has no APIs to give Key_0 to the off-chain server and opens no TCP port except 443 for HTTPS. Therefore, the attacker compromises the key storage is difficult. If the attacker cannot get Key_0, then the attacker has no clue to find password from Key_2, of which encrypted form is found in the off-chain server. It is the reason because the attacker does not know two variables, Key_0 and Key_1 with one equation ($Key\_0 = Key\_1 \oplus Key\_2$). Therefore, attacker must compromise the key storage. If the attacker would compromise the key storage, the attacker could get the encrypted Key_0, encrypted serial number from table in Fig. 3 and could get the artwork serial number from table in Fig. 2. The attacker launches the brute force attack or dictionary attack on encrypted Key_0 to know the administrator password. The attacker decrypts encrypted Key_0 and gets the result Key_0. The attacker has to know Key_0 is really Key_0. To do so, the attacker has to decrypt encrypted serial number, that is random number. The attacker has to retrieve serial number with the decrypted serial number, but the attacker does not know whether retrieved serial number is really true serial number. Because wrong serial number can be match with other serial number coincidentally.

Moreover, the administrator's password for Key_2 and the administrator's password for Key_0 are different. There is no verifiable value to check whether guessed password for Key_2 is right or not. In sum, there is no verifiable value to check whether guessed password for Key_2 and Key_0 is right or not. Therefore, our system is very secure.

5 Discussion and Contribution

The artist puts together his/her artwork list in the excel or MS word not systematically. Maybe the computer of the artist is out of work or the list accidentally is deleted or not well managed. Therefore, the artist wants systematically service to keep his/her artwork list. Our system provides that service with blockchain.

The artist can track his/her artwork because the list of owners of his/her artwork is stored in blockchain and other information in the off-chain server.

Proposed system does not use tokenization of artwork ownership. However, the proposed system can track the transferring ownership. If the payment process is included in the transferring ownership process, and the proposed system charges some percentage of the payment, e.g., 3%, then the proposed system can give back some percentage of payment, e.g., 1% to the artist. It is a big advantage and an attractive service for the artist.

The buyer can verify the authenticity of the artwork with our system. Our system improves the certainty of authenticity with QR code and password. For instance, a thief must steal the artwork and the certificate of authenticity together. Even if the thief successfully steals the artwork and the certificate of authenticity, the thief does not know the password. Therefore, the thief cannot prove the ownership and the authenticity of the artwork to the buyer. Therefore, our system improves the certainty of ownership and authenticity of artwork.

Another case, the bad guy sells the forgery artwork and the copy of the certificate of authenticity, the buyer cannot easily detect it is forgery artwork. With our system, the bad guy wants to sell the forgery artwork and the copy of the certificate of authenticity, but he does not know the password. Therefore, the buyer can easily detect whether it is something wrong with our system.

Think that owner tries to sell the forgery artwork and genuine certificate of authenticity and keep genuine artwork. The owner can sell the forgery artwork and genuine certificate of authenticity. However, the owner cannot sell again genuine artwork forever, even if the owner has the copy of the certificate of authenticity, because the password (Key_1) is changed and the owner does not know new owner's password. Also, all history of owner is stored in the distributed ledger of blockchain, so if the forgery is revealed, the owner who sold the forgery artwork can easily be revealed.

Thereby, our system bridges the "blockchain air gap". There is no solution to solve the blockchain air gap problem, but our system solves this problem efficiently and is the first novel solutions.

Our system requires KYC. Our system is publicly accessible, so anyone can upload an artwork in our system. However, with KYC, if one uploads forgery artwork or wrong information, then easily tracks who did it. It also mediates the first mile problem, that is uploading wrong information in the blockchain which cannot change written information in the distributed ledger.

Our system is developed for art domain. However, some domain, that the verification of ownership and authenticity are important, can adapt our system. For example, in the luxury market, there are a lot of fake luxury goods. This makes brand image bad and decreases sales of the luxury brand company. Moreover, in the second hands luxury market, there is no method to check whether the goods is fake or real luxury goods. Our system can be a good candidate for verifying ownership and authenticity. For the luxury market, the QR code can be printed in the tag of luxury goods.

Another example is bike market because the bike is easily robbed and the number of robberies is extremely high. For this case, the ownership proof is particularly important. Our system can be adapted for this field.

6 Conclusions

Blockchain can provide the integrity of the record written in distributed ledger, so for the authenticity of artwork the blockchain is an attractive system. This paper proposed a novel blockchain based authenticity system providing the authenticity of artwork based on the certificate of authenticity. The proposed system provides more certainty of the authenticity of artwork compared with existing certificate of authenticity. In the proposed system, QR code, i.e., the encrypted serial number is used for verifying the ownership and authenticity of artwork. The related password of QR code is changeable, but printed QR code is not required to be changed. In our system, the password is not stored in any system, so the attacker cannot get the password even if the system is compromised.

This paper tries to address the blockchain air gap efficiently but not expensively. Therefore, the artwork and the certificate of authenticity will be sold together. This paper is the first trial to address the blockchain air gap.

Acknowledgements This paper was supported by the Konyang University Research Fund in 2020.

References

1. Verisart. https://verisart.com. Cited 16 May 2021
2. Artory. https://www.artory.com. Cited 16 May 2021

3. Codex Protocol. https://www.codexprotocol.com. Cited 16 May 2021
4. Whitaker, A., Kraeussl, R.: Blockchain, fractional ownership, and the future of creative work. Goethe University Center for Financial Studies, Working Paper No. 594
5. Whitaker, A., Bracegirdle, A., de Menil, S., Gitlitz, M.A., Saltos, L.: Art, antiquities, and blockchain: new approaches to the restitution of cultural heritage. Int. J. Cult. Policy 1–18 (2020)
6. Evans, T.J.: Cryptokitties, cryptography, and copyright. Am. Intellect. Property Law Assoc. Q. J. **47**, 219–247 (2019)
7. Whitaker, A.: Art and blockchain: a primer, history, and taxonomy of blockchain use cases in the arts. J. Entrepreneurship Arts **8**(2), 21–46 (2019)
8. O'Dwyer, R.: Limited edition: producing artificial scarcity for digital art on the blockchain and its implications for cultural industries. Convergence **26**(4), 874–894 (2020)
9. Eden, T.: How I became Leonardo da Vinci of the blockchain. https://shkspr.mobi/blog/2018/06/how-i-became-leonardo-da-vinci-on-the-blockchain. Cited 16 May 2021
10. Schneider, T.: Cryptocurrencies explained. Artnet News. https://news.artnet.com/art-world/cryptocurrencies-explained-part-three-1248863. Cited 16 May 2021

Image Steganography Using GANs

Ketan Ramaneti, Pranavi Kakani, Chaitanya Krishna, and Sujatha Rajkumar

Abstract Secure transmission of information across networks is of esteem importance to meet confidentiality and privacy goals and therefore is a huge concern to modern society. To prevent information from being compromised during communication, many steganography or data hiding algorithms have been introduced in recent years. Steganography refers to hiding secret information into cover files like text, images, audio etc., to secure communications. The term Image Steganography is used to describe steganography processes where images are used as cover files. Conventional steganography algorithms are observed to impose high risk of being deciphered by intelligent systems. This paper proposes a system that eliminates such a risk by hiding secret information inside stego images generated using Generative Adversarial Networks (GANs) and then safely reproducing it using an Extractor model. The experimental results of the proposed system demonstrated an accuracy of 92.34% achieved by the extractor model while retrieving the hidden information from the stego images and an overall steganography capacity of 93.75e−3 bytes/pixel for the entire system. Thus, the proposed system reduces the probability of any external intelligent system intercepting the communication and stealing information, like in the case of systems following traditional algorithms.

Keywords Generative adversarial networks (GANs) · Image steganography · Steganography capacity · Convolution neural networks (CNN) · Secure communication

K. Ramaneti (✉) · P. Kakani · C. Krishna · S. Rajkumar
School of Electronics Engineering, Vellore Institute of Technology, Vellore, India
e-mail: ramanetiketan@gmail.com

P. Kakani
e-mail: pranavikakani2001@gmail.com

C. Krishna
e-mail: chaitanya.krishna1002@gmail.com

S. Rajkumar
e-mail: sujatha.r@vit.ac.in

R. Lee (ed.), *Computer and Information Science 2021—Summer*,
Studies in Computational Intelligence 985,
https://doi.org/10.1007/978-3-030-79474-3_12

1 Introduction

The practice of securing information by minimizing risks of it being compromised is known as information security. With reference to information security concepts, cryptography is a method that makes use of secret codes to make the data unreadable or encrypt the meaning of data for protecting information and communications. Steganography on the other hand is a method used to hide the entire existence of data by embedding it inside another file of similar type known as a cover file. Various types of files like text, image, audio, video etc., can be used as cover files for steganography, thus categorizing the process into different types. For example, the term Image Steganography is used to group steganography algorithms that use images as cover files. Recently, research on image steganography has demonstrated great potential for using GAN and neural networks [1].

GANs are deep learning based generative models proposed by Ian Goodfellow in 2014. GANs have proven to be quite effective in many fields since it was proposed. The architecture of the GAN model involves two sub models: a Generator model for generating new data and a Discriminator model for classifying the new generated data as real i.e. obtained from the domain or fake data generated by the generator model. GANs are rooted in a game theoretic scenario in which the generator model competes against the adversary model. GANs have a lot of applications from improving cybersecurity by preventing adversarial attacks and anonymizing data to privacy preservation to creating avatars and more [2–4]. Although there is much to explore, they also have the ability to generate high quality data and produce sharper, discrete outputs than other algorithms. One recent advancement for GANs is its ability to produce efficient results when used for hiding data or steganography [5–7].

Steganography can be classified into two main categories, Irreversible Data Hiding (IDH) and Reversible Data Hiding (RDH) [8, 9]. The process of retrieval of the secret hidden data on the receiver end is called Steganalysis. Image Steganalysis algorithms are developed to determine whether an image is embedded based on statistical features of distortion. Finding patterns in the data, analyzing the data can be done with ease using Artificial Intelligence (AI). Applications of AI in IDH have gained a lot of attention and breakthroughs have been achieved in these methods and performances. Adaptive steganography algorithms are proposed to minimize the embedding distortion; however, the distortions can also be detected by using advanced machine learning based steganalysis algorithms [10]. To avoid this risk of being decrypted by steganalysers, a concept called Steganography Without Embedding (SWE) was proposed.

Digital images contain information in the form of pixel brightness, color, texture, edges, contours and pixels themselves. With appropriate processing of this information, a mapping relationship can be created between Image information and the secret information to be hidden [11]. The requirements of such mappings are suitable digital images that can carry information via mapping and the algorithm for performing the mapping. In recent years, several SWE methods have been proposed. Some of these works have proposed SWE methods to map the secret information from the semantic features of the images with the help of bag of words, image hashing, a class of texture synthesis which involves reversible mathematical functions. But these methods have limited capacity. Therefore, in this paper we propose an image steganography technique utilizing GANs for hiding secret data in order to increase the steganography capacity and secure the entire communication process.

2 Proposed System

We propose a three-phase steganography framework, as shown in Fig. 1 that generates stego images containing secret data which is indecipherable by a third person other than the sender and receiver. A total of three Neural Networks are used in the proposed framework.

In the first phase, a DCGAN is trained using The Street View House Numbers (SVHN) Dataset and the generator G is obtained from the optimized DCGAN, which is capable of generating new stego images based on the input noise vector z. In the second phase, a CNN network-based Extractor, E is trained to extract the hidden noise vector from the stego images, based on the recovery errors from a large number of noise vectors and generated images from G. In the third phase, the secured data transmission is implemented and the results were analyzed. The sender and receiver

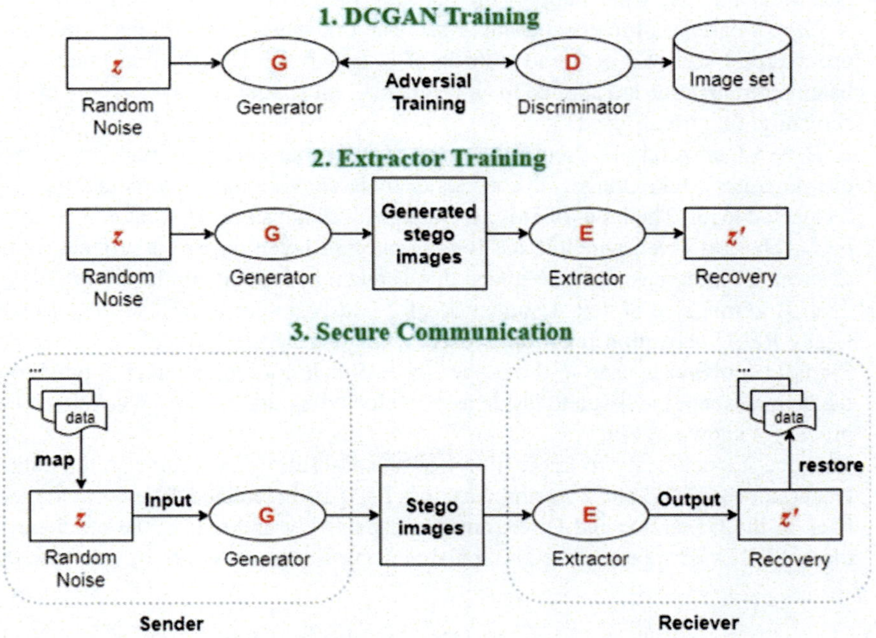

Fig. 1 Phase-wise block diagram for the proposed system

are given the unique Generator and Extractor respectively. Before passing data to the generator, the secret data is divided into several groups of 3 bits each and each bit is mapped to a certain interval between −1 and 1, which is used to define a noise vector z corresponding to the data. The mapped noise z is passed through G and a stego image is generated. This image is transmitted to the receiver. After the receiver receives the stego image, he passes it through E to extract the vector z′ which is similar to that of the original vector z. The extracted noise vector z′ is reverse mapped based on the mapping mechanism described, to obtain the secret data hidden in the stego images. Phases of the architecture are further explained in this section. A structured phase-wise diagram for the same is shown in Fig. 1.

2.1 DCGAN Training

In this phase, a Generator G and a Discriminator D are trained using the Street View House Numbers (SVHN) dataset. The dataset consists of color images of house numbers taken from Google street view. G is a transpose convolution network that consists of one fully connected layer and three transpose convolution layers and is used to fit the real images' data distribution in a training set to produce a fake image. Transpose convolution layers have double the width and height and halve the depth of the previous layer. The input fed to G is a noise vector 'z' of size 1×256, which later we obtain using the mapping method. The output of G is a fake image of size $32 \times 32 \times 3$ based on the given training set. The transpose convolutions in this paper use a kernel size of 4×4 and a stride of 2. Leaky ReLU activation function and batch normalization are applied to all layers in G except the last layer, where the tanh activation function is used.

D is a Convolutional Neural Network (CNN) that compares real images with the generator's fake images. D consists of three convolutional layers and one fully connected layer. The input of D is an image with dimensions $32 \times 32 \times 3$ produced by G. The last layer, which is the fully connected layer, outputs a probability that determines whether or not the image that is taken as input is produced by G i.e., is fake. A kernel size of 4×4 and stride of 2 is used in convolutions for D as well. Leaky ReLU activation function is used in the three convolutional layers whereas Sigmoid function is used in the output layer. Batch normalization is applied to all the layers except the first and last layers. A block diagram for the DCGAN Training process is shown in Fig. 2.

Since it's necessary to know how well the algorithm is working on the dataset, Generator loss L(G) and Discriminator loss L(D) are calculated. They provide feedback to the Generator and Discriminator respectively and realize the efficiency of the model. Discriminator loss is calculated using Binary cross-entropy defined by:

$$L(D) = \frac{1}{m}\sum_{i=1}^{m}\log(D(x^i)) + \log(1 - D(G(z^i)))$$

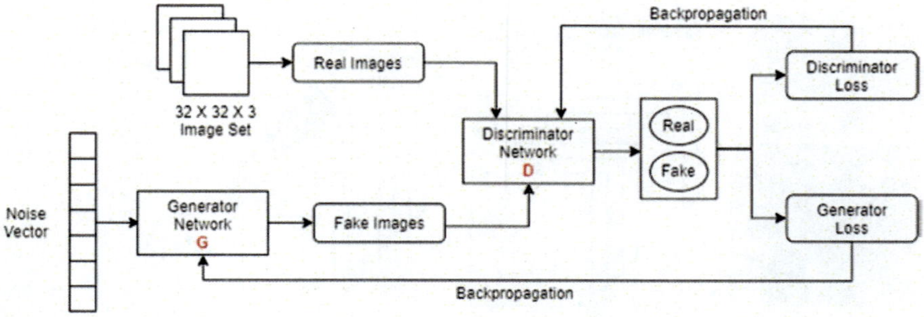

Fig. 2 DCGAN training

D aims to give an output of 1 for real images from the training set and 0 for fake images produced by G. The objective of G is to deceive D into classifying fake images as real and real images as fake, thus producing higher values of L(D). Therefore, the higher the L(D), the higher will be the credibility of the proposed model. However, it is important to note that L(D) and L(G) must attain equilibrium. The G loss is the same as the D loss, except the labels are flipped. The equation for the L(G) is given below:

$$L(G) = \frac{1}{m}\sum_{i=1}^{m}\log\left(1 - D\left(G\left(z^i\right)\right)\right)$$

2.2 Extractor Training

Extractor E is a CNN model aimed to recover the secret hidden data from the images generated by the Generator. For accurate extraction of the hidden data, Extractor should follow the same principle as that of the Discriminator which was used as a feedback mechanism for Generator training and should be the exact opposite of the Generator which generates images from noise vectors. For that, while designing E, we followed the principle of Discriminator with convolution layers followed by fully connected layers.

The input of E is a 32 × 32 × 3 fake image generated by G and the output is a noise vector z′ with a dimension of 1 × 256. The aim of E is to ensure that the generated output z′ is identical to the noise vector z corresponding to the fake image which is given as input. This is similar to an encoder in an autoencoder model which maps an image to a certain vector. So, the convolution layers of E are similar to that of a standard encoder model. The first convolution layer consists of 32 channels with kernel size of 3 × 3, stride 2 and ReLU activation function that outputs a vector with dimensions 15 × 15 × 32. The second convolution layer is similar to the first convolution layer with 16 channels which outputs a vector with dimensions 7 × 7 ×

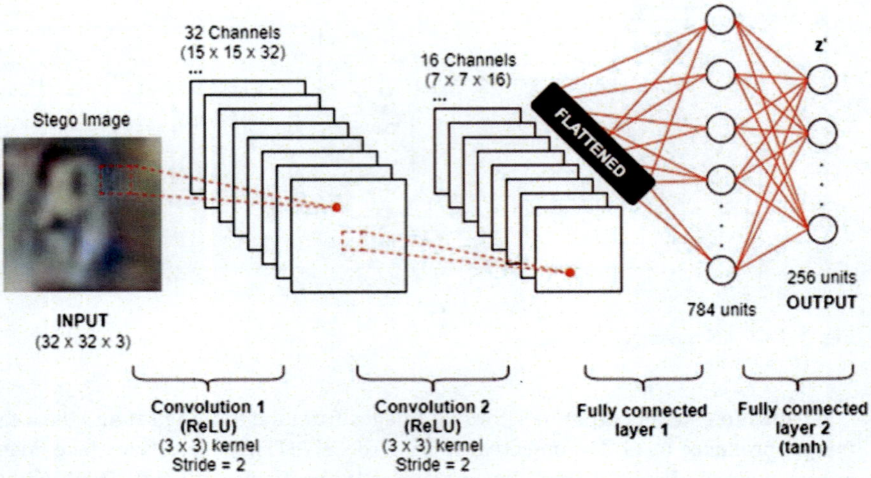

Fig. 3 Extractor architecture

16. The next layer is a fully connected layer followed by the output layer. The output layer of E consists of 256 units to obtain the z′ vector. Since the noise vector always lies between −1 to 1, tanh activation function is used at the output layer of E. The weight parameters in each layer are learned to fit the nonlinear mapping between the input stego images and the output noise vector. To enhance these nonlinear learning capabilities, we used appropriate activation functions, the ReLU and tanh functions as mentioned, dropout operation was introduced in the last convolution layers. Batch Normalization was applied to each layer. The architecture of E is illustrated in Fig. 3.

For training E, random noise vectors, z with size 1 × 256 and range between −1 and 1 were generated and passed through G to generate stego images. These noise vectors are independent of the noise vectors used while training the generator. This prevents the problem of overfitting due to certain noise vectors. The generated stego images are passed as inputs to the extractor for training. E outputs z′ which is the recovered noise vector identical to z. The optimization goal while training E is to minimize the deviation between the original noise vectors z and the recovered vectors z', extracted from the stego images. The deviation between z and z′ is calculated using the loss function of the extractor which is defined as the mean square loss given by the equation:

$$L(E) = \frac{1}{m}\sum_{i=1}^{m}(z - E(G(z)))^2 = \frac{1}{m}\sum_{i=1}^{m}(z - z')^2$$

Here, m is the size of the training batch, z is the original noise vector, G is the generator, E is the Extractor, z′ is the extracted noise vector from E, G(z) is the stego image generated by G from z and z′ = E(stego) = E(G(z)).

E is trained for a specific G which is obtained from the DCGAN training mentioned above. The aim of Extractor is to minimize the loss, L(E). L(E) is used as a feedback mechanism and the weights are updated using backpropagation. When the loss is small which implies the network E is convergent, E is used to recover secret data from the stego images generated by G. As E is trained solely using the images produced by G, it will be impossible to replicate it and extract the hidden information from the stego images unless the intercepting third party knows about G.

2.3 Secure Communication

In this phase, G and E are utilized to establish a secured communication between the sender and receiver. The sender is provided with the unique Generator G and the receiver is provided with the corresponding unique Extractor E. As shown in Fig. 1, before generating stego images via G, the noise vector z is produced from the secret data using a mapping method realized in [12]. The mapping method maps secret binary bits to noise vectors. Firstly, the binary bits are divided into various groups of 3 bits each. For example, the binary bits {100101010} are divided into {100}, {101}, {010}. Following this, each group is mapped to a random noise within a particular interval according to the equation:

$$r = random\left(\frac{m}{2^{k-1}} - 1 + \delta,\ \frac{m+1}{2^{k-1}} - 1 - \delta\right)$$

where k denotes the number of bits in the group to be mapped which is 3 in our case, m denotes the decimal value of the group, and Delta(symbol) denotes the gap between the divided intervals. The minimum and maximum values of any interval are -1 and 1 respectively. A noise vector is finally generated using the mapping method. This method is shared by both the sender and the receiver. On the sender's side, the secret data is mapped to the noise vector whereas on the receiver's side, the secret data is retrieved from the noise vector. The mapping groups and their corresponding intervals are shown in Table 1.

After mapping the data and producing the noise vector z, it is passed through G from the sender's side to generate a stego image that contains the secret data and is indecipherable by any mechanism other than the Extractor, E. The stego image is transmitted to the receiver. The receiver receives the stego image and passes it through E to extract the noise vector z′ which is similar to z. The extracted noise vector z′ is reverse mapped according to the mapping method described and the secret data is retrieved by the receiver. Thus, a secured transmission of data has been established between the sender and the receiver.

The proposed model requires the size of the noise vector, z to be 256 for the generator to generate stego images and the extractor can recover the noise vector

Table 1 Mapping from groups to intervals

Group	Interval
000	(−0.999, −0.751)
001	(−0.749, −0.501)
010	(−0.499, −0.251)
011	(−0.249, −0.001)
100	(0.001, 0.249)
101	(0.251, 0.499)
110	(0.501, 0.749)
111	(0.751, 0.999)

containing 256 values. As described in the mapping method, each value of z corresponds to 3 bits of information and thus, 768 bits of information is required to map and produce a noise vector of 256 values.

If the user wants to encrypt a bit-string of size less than 768 bits, a noise vector containing 256 values can't be produced and thus G can't generate a stego image for secure transmission. To avoid such cases, first the length of the information in bits is calculated and stored (*L*). If L is smaller than 768, *768-L* random binary bits are concatenated to the information at the MSB to make the information 768 bits in size. The bit-string is now mapped to produce a noise vector containing 256 values and the noise vector is passed through After retrieving the secret information via extractor, L bits starting from LSB are considered as the actual secret information and the remaining 768-L bits, which are random bits, are excluded.

Similarly, if the user wants to encrypt a bit-string of size greater than 768 bits, multiple noise vectors should be produced and the corresponding stego images should be generated by G. When L is greater than 768, L is divided by 768 and the quotient q, and remainder r, are stored. The information is divided into q labelled subgroups such that each subgroup consists of 768 bits and the order of subgroups is noted. The remaining r bits are stored in another subgroup. Therefore, a total of q + 1 subgroups are formed. Since r is always less than 768, the last subgroup follows the above proposed principle where information size is less than 768 bits and the bit-string in this subgroup is made to have 768 bits. For secure transmission, each subgroup is mapped to produce the corresponding noise vectors, and passed through G to generate q + 1 stego images and the corresponding data is extracted using extractor. Since the order of the subgroups is known, ultimately the order of the stego images can be derived and the secret information can be retrieved by reverse mapping the extracted noise vector and concatenating the bits in its respective order.

3 Experimental Results

3.1 DCGAN Training Results

The training of DCGAN is carried out on the SVHN dataset. The input of the Generator is a noise vector z, of dimension 1 × 256. We set the learning rate to 0.0002 and trained the architecture for 50 epochs. The losses obtained for the Generator and Discriminator are shown in Fig. 4.

The output of the Generator is a set of images each of dimension 32 × 32 × 3. A sample output of 8 images from the Generator used in the proposed model is shown in Fig. 5.

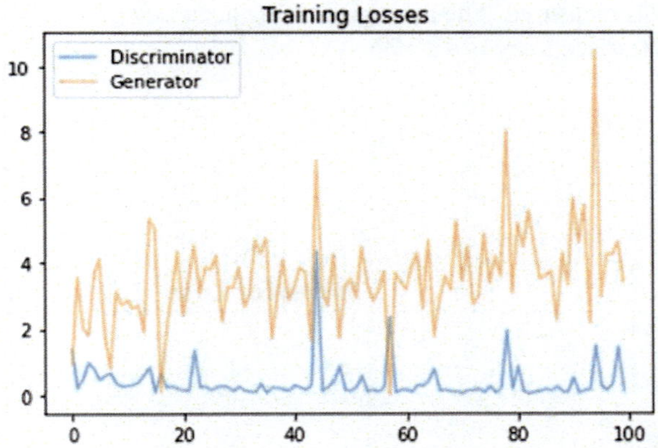

Fig. 4 Generator and discriminator losses

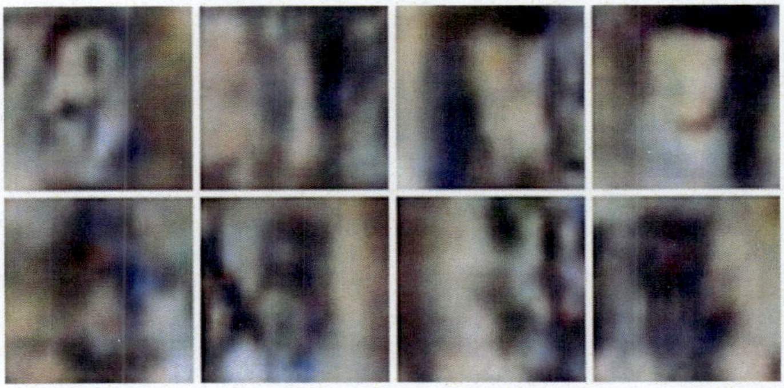

Fig. 5 Generator sample output

3.2 Extractor Training Results

Around 30,000 images were generated using G and the corresponding noise vectors were stored and marked with their respective images. These images and noise vectors were split into train and test sets in the ratio of 0.85:0.15 for training and testing E. In the training procedure, minibatch was set to a size of 128, Stochastic Gradient Descent optimizer with a learning rate equal to 0.01 and momentum equal to 0.9 was used. The Extractor was trained for 150,000 steps. The values of loss function of E after each step are shown in Fig. 6.

It can be observed that the value of loss function decreases slowly for the first 20,000 steps and decreases rapidly after 20,000 steps eventually converging to a small value which implies that the output of Extractor is similar to the target noise vectors. For testing the Extractor, the accuracy of extraction of the secret data from the test set was measured. The average training accuracy of the Extractor is 96.67%. The average test accuracy is 92.34%.

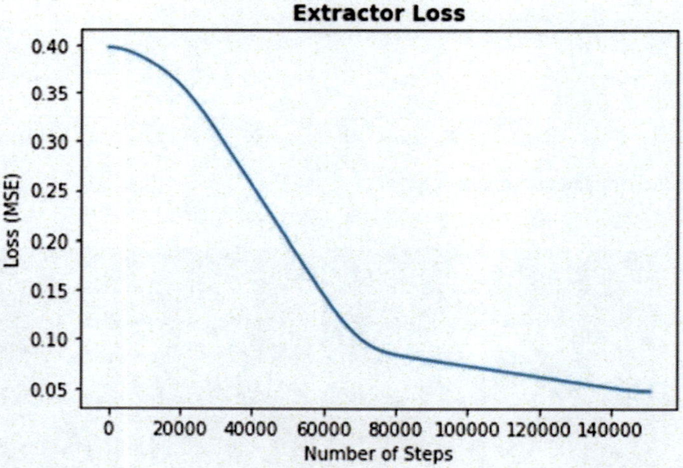

Fig. 6 Extractor loss

3.3 Secure Communication Implementation

First, the secret information in bits is mapped to produce the corresponding noise vector according to the described mapping method, a part of which is shown in Fig. 7.

The produced noise vector is passed through G to generate the corresponding stego image, as shown in Fig. 8.

This generated image securely stores the secret data and is transmitted by the sender. The receiver receives this stego image and passes it through E to extract the corresponding noise vector, a part of which is shown in Fig. 9.

The noise vector is reverse mapped according to the mapping method to retrieve the secret data in bits.

```
mapping_method(secret_data)

Secret data:
11001101000100111011000010010001011001110011101110000011000101110100010100111110011001010011101
Secret data divided into sub-groups of length 3:
['110', '011', '010', '001', '001', '110', '110', '000', '100', '100', '010', '110', '011', '1
Mapped noise vector:
[0.599, -0.157, -0.356, -0.63, -0.674, 0.731, 0.509, -0.076, 0.114, 0.033, -0.282, 0.533, -0.1
```

Fig. 7 A part of the secret data and the corresponding mapped noise vector

```
import matplotlib

stego_image = G(z)

for img1 in stego_image:
  img1 = img1.detach().cpu().numpy()
  img1 = np.transpose(img1, (1, 2, 0))
  img1 = ((img1 +1)*255 / (2)).astype(np.uint8)
  img1 = img1.reshape((32,32,3))
  plt.figure(figsize = (16,4))
  im1 = plt.imshow(img1)
```

```
/usr/local/lib/python3.7/dist-packages/torch/nn/functional.py:1698: UserWarning: nn.functional.
  warnings.warn("nn.functional.tanh is deprecated. Use torch.tanh instead.")
```

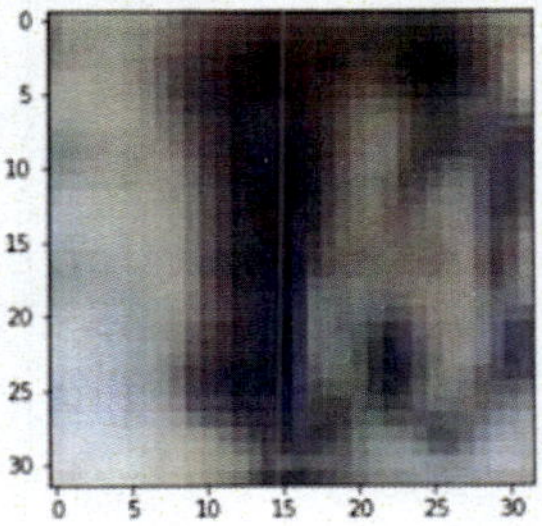

Fig. 8 Stego image generated by G from the mapped noise vector

```
extracted_z = E.predict(stego_image)

print(extracted_z)

[0.618, -0.153, -0.352, -0.611, -0.658, 0.751, 0.516, -0.069, 0.118, 0.051, -0.268, 0.536,
```

Fig. 9 A part of the retrieved noise vector from the stego image

3.4 *Capacity*

Absolute capacity (bytes/image) for a steganographic system is calculated as the number of bytes stored in a single image, and the Relative capacity (bytes/pixel) for the same is calculated as the value of Absolute capacity by the image size in pixels. Thus, the number of bytes stored should be as high as possible and the stego image size be as low as possible to maximize capacity. As 32 × 32 RGB images are of standard minimal sizes, and quality of the stego image is not of priority in the proposed method, we decided to train the generator to generate stego images of size 32 × 32 pixels. As the number of bytes stored in a stego image is directly proportional to the size of z, it was realized that with progressive increase in the size of z, the extractor accuracy drops as shown in Fig. 10.

The comparison was drawn between the z vector of size 64, 128, 256, 512, 1024 and the corresponding extractor's test accuracy which was found out to be 96.34%, 95.63%, 92.34%, 82.58%, 68.63% respectively. An optimum value for the number of

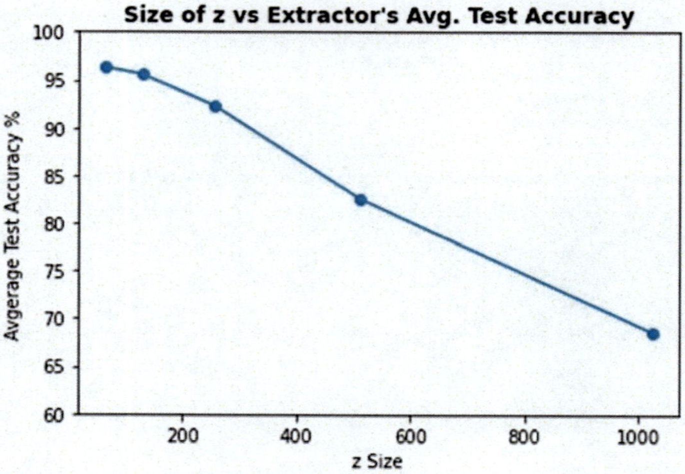

Fig. 10 Drop in extractor accuracy with increasing z size

Table 2 Steganography capacities of SWE methods

Method	Absolute capacity (bytes/image)	Image size	Relative capacity (bytes/pixel)
[12]	37.5	64 × 64	9.16e−3
[13]	1.125	512 × 512	4.77e−6
[14]	3.72	≥512 × 512	1.42e−5
[15]	2.25	512 × 512	8.58e−6
[16]	25–100	480 × 640	8.14e−5 to 3.26e−4
[17]	64 × 64	800 × 800	6.40e−3
[18]	1535–4300	1024 × 1024	1.46e−3 to 4.10e−3
Ours	96	32 × 32	93.75e−3

bytes to be stored (z size equal to 256) was chosen so the extractor accuracy doesn't get compromised.

As described, a bit string of 768 bits can be securely embedded in a generated stego image of size 32 × 32 pixels in the proposed system. Implies, the capacity of the proposed method is 96 bytes/image or 0.09375 bytes/pixel.

The steganography capacities of SWE methods are lower than traditional embedding based steganography. The proposed method in this paper is a new SWE method. The capacity comparison of various SWE steganography methods is shown in Table 2.

4 Conclusions

In this paper, a three-phase steganography framework using DCGANs has been proposed. In the first phase, a DCGAN model was used to train the Generator and Discriminator with the SVHN dataset. The trained model generated images in such a way that the discriminator could not classify them any differently than the actual training set images. In the second phase, an Extractor was trained to extract the hidden noise vector from the stego images based on the recovery errors of a large number of noise vectors and images produced by G. The main objective of the paper was served in the last phase, where secure data transmission was performed. The secret data that were to be transmitted were mapped to a noise vector z using the mapping technique and hidden inside the stego images produced by G. These images after being transmitted across the network, were fed to the Extractor where the secret data was safely retrieved. The proposed system was observed to be highly effective in the experiments performed. The extractor model achieved a very good accuracy of 92.34% for retrieving the hidden information from the stego images, making it reliable, and an overall capacity of 93.75e−3 bytes/pixel was noted for the entire system. The observed steganography capacity of the proposed model was higher than that of standard SWE methods. These experimental results of the proposed model demonstrate that it greatly improves the imperceptibility of secret data and its safe retrieval on the receiver end, thus ensuring secure communication across networks.

References

1. Liu, J., et al.: Recent advances of image steganography with generative adversarial networks. IEEE Access **8**, 60575–60597 (2020). https://doi.org/10.1109/ACCESS.2020.2983175
2. Cachin, C.: An information-theoretic model for steganography. In: Information Hiding, pp. 306–318. Springer, Berlin, Germany (1998). [Online]. Available: https://doi.org/10.1007/3-540-49380-8_21
3. Reed, S., Akata, Z., Yan, X., Logeswaran, L., Schiele, B., Lee, H.: Generative adversarial text to image synthesis (2016), [Online]. Available: https://arxiv.org/abs/1605.05396
4. Volkhonskiy, D., Borisenko, B., Burnaev, E.: Steganographic generative adversarial networks. CoRR abs/1703.05502 (2017), [Online]. Available: http://arxiv.org/abs/1703.05502
5. Liu, M.-M., Zhang, M.-Q., Liu, J., Gao, P.-X., Zhang, Y.-N.: Coverless information hiding based on generative adversarial networks. Yingyong Kexue Xuebao/J. Appl. Sci. **36**, 371–382 (2018). https://doi.org/10.3969/j.issn.0255-8297.2018.02.015
6. Li, J., Niu, K., Liao, L., Wang, L., Liu, J., Lei, Y., Zhang, M.: A generative steganography method based on WGAN-GP (2020). https://doi.org/10.1007/978-981-15-8083-3_34
7. Zhang, Z., et al.: Generative steganography by sampling. IEEE Access **7**, 118586–118597 (2019). https://doi.org/10.1109/ACCESS.2019.2920313
8. Zhang, Z., Fu, G., Di, F., Li, C., Liu, J.: Generative reversible data hiding by image-to-image translation via GANs. Secur. Commun. Networks **2019**, 1–10 (2019). https://doi.org/10.1155/2019/4932782
9. Sarkar, T., Sanyal, S.: Reversible and irreversible data hiding technique (2014)
10. Pibre, L., Jérôme, P., Ienco, D., Chaumont, M.: Deep learning is a good steganalysis tool when embedding key is reused for different images, even if there is a cover source-mismatch (2015), [Online]. Available: https://arxiv.org/abs/1511.04855
11. Ke, Y., Zhang, M., Liu, J., Su, T., Yang, X.: Generative steganography with Kerckhoffs' principle based on generative adversarial networks (2018). https://arxiv.org/abs/1711.04916
12. Hu, D., Wang, L., Jiang, W., Zheng, S., Li, B.: A novel image steganography method via deep convolutional generative adversarial networks. IEEE Access **6**, 38303–38314 (2018). https://doi.org/10.1109/ACCESS.2018.2852771
13. Zhou, Z., Sun, H., Harit, R., Chen, X., Sun, X.: Coverless image steganography without embedding. In: Proceedings of International Conference on Cloud Computing Security, pp. 123–132 (2015)
14. Zhou, Z.-L., Cao, Y., Sun, X.-M.: Coverless information hiding based on bag-of-words model of image. J. Appl. Sci. **34**(5), 527–536 (2016)
15. Zheng, S., Wang, L., Ling, B., Hu, D.: Coverless information hiding based on robust image hashing. In: Intelligent Computing Methodologies, pp. 536–547. Springer, Cham, Switzerland (2017). https://doi.org/10.1007/978-3-319-63315-2_47
16. Otori, H., Kuriyama, S.: Texture synthesis for mobile data communications. IEEE Comput. Graph. Appl. **29**(6), 74–81 (2009)
17. Xu, J., et al.: Hidden message in a deformation-based texture. Vis. Comput. Int. J. Comput. Graph. **31**(12), 1653–1669 (2015)
18. Wu, K.-C., Wang, C.-M.: Steganography using reversible texture synthesis. IEEE Trans. Image Process. **24**(1), 130–139 (2015)

Coverage-Guided Fairness Testing

Daniel Perez Morales, Takashi Kitamura, and Shingo Takada

Abstract Software testing is a crucial task. Unlike conventional software, AI software that uses decision-making algorithms or classifiers needs to be tested for discrimination or bias. Such bias can cause discrimination towards certain individuals based on their protected attributes, such as race, gender or nationality. It is a major concern to have discrimination as an unintended behavior. Previous work tested for discrimination randomly, which has resulted in variations in the results for each test execution. These varying results indicate that, for each test execution, there is discrimination that is not found. Even though it is nearly impossible to find all discrimination unless we check all possible combinations in the system, it is important to detect as much discrimination as possible. We thus propose Coverage-Guided Fairness Testing (CGFT). CGFT leverages combinatorial testing to generate an evenly-distributed test suite. We evaluated CGFT with two different datasets, creating three models with each. The results show an improvement in the number of unfairness found using CGFT compared to previous work.

Keywords Fairness · Testing · Machine learning · Combinatorial testing

1 Introduction

Recent advances in machine learning (ML) techniques make predictions and decisions with higher precision and accuracy, resulting in a wider use of this type of software. There is a wide variety of ML applications, such as product recommenda-

D. Perez Morales (✉) · S. Takada
Keio University, Yokohama, Japan
e-mail: daniel.morales@keio.jp

S. Takada
e-mail: michigan@ics.keio.ac.jp

T. Kitamura
National Institute of Advanced Industrial Science (AIST), Tokyo, Japan
e-mail: t.kitamura@aist.go.jp

R. Lee (ed.), *Computer and Information Science 2021—Summer*,
Studies in Computational Intelligence 985,
https://doi.org/10.1007/978-3-030-79474-3_13

tions [20], deciding who will be granted a loan [22], self-driving car systems [11], medical diagnosis and treatment [25] and sentencing criminals [9].

As such decision-making software is becoming more ubiquitous in our daily life, there is a growing concern whether these decisions are biased. Discrimination injected in the COMPAS, which is used in U.S. courts to assist in sentencing, is one of many examples of such concerns. COMPAS computes risk-assessment scores for suspected criminals, which are an estimated probability that the person arrested for a crime is likely to commit another crime [10]. Despite that these scores play a critical role in the criminal justice system process, the software is known to include bias: this software is "particularly likely to falsely flag black defendants as future criminals, wrongly labeling them this way at almost twice the rate as white defendants; white defendants were mislabeled as low risk more often than black defendants" [13]. Another example of discrimination in software is a classifier created by Amazon used in their recruiting process, which was proven to discriminate against women [7]. These models formed from machine learning are based on the training dataset. Unfortunately, such dataset is not always complete or may include bias, resulting in the model to have unintended bias. Such bias needs to be found.

Testing is one approach to solving the software fairness issue. Software testing analyzes software dynamically, by executing it with test cases, for the purpose of detecting bugs or assuring some quality based on coverage criteria. Fairness testing thus can analyze software by executing test cases, to detect discrimination. Testing software fairness has been under active investigation, resulting in various findings such as finer classification of software fairness (e.g., group and individual fairness [10]), and efficient algorithms for test generation [1, 27].

AEQUITAS [27] is a state-of-the-art black-box testing method for software fairness, targeting individual fairness. As previous black-box testing techniques such as THEMIS [10] largely rely on random sampling for generating test cases, AEQUITAS devises a search strategy in generating test cases in order to find discrimination more effectively. Specifically, AEQUITAS's so-called *directed strategy* structures its test case generation into two phases of *global search* and *local search*. While the global search still uses random sampling to find discrimination, the local search generates test cases by applying a strategy, called *perturbation*, to the discriminations detected in the global search to find more discrimination. Experiments show that AEQUITAS can detect more cases of individual discrimination than other existing techniques.

In this paper, we propose a fairness testing technique, called *Coverage-Guided Fairness Testing (CGFT)*, by refining AEQUITAS with the concept of coverage criteria of software testing. CGFT introduces the concept of combinatorial t-way testing [18], a testing technique based on *t-way coverage criteria*, in the global search of AEQUITAS. The aim behind this attempt is to leverage previous findings that t-way testing tends to sample higher quality data than random testing, in terms of diversity of sampled data and bug detecting capability [28]. Our hypothesis is that CGFT improves AEQUITAS in finding more discriminations at a reasonable execution cost.

The rest of this paper is organized as follows. Section 2 provides a background of our work. Section 3 shows the details of CGFT. In Sect. 4, we evaluate CGFT. In Sect. 5, we mention the main threats to validity. Section 6 presents related work. Section 7 concludes the paper and mentions future work.

2 Background

2.1 Individual Discrimination

Our research is focused on testing ML classifiers looking for discrimination. Like other studies on fairness testing [1, 10, 27], the software we target in this paper handles human data in a format which consists of a sequence of values that are assigned to attributes (a.k.a., parameters). Each data identifies a human. For example, the LOAN software in [10] is used to decide whether to give a loan to an applicant depending on their attributes, such as 'name', 'age', 'race', 'income', 'savings', 'employment status', and 'requested loan amount'.

In general terms, discrimination can be understood as the lack of fairness, that is, the presence of bias. In the area of software fairness, two main kinds of software discriminations have been recognized in the literature [10]: *group discrimination* and *individual discrimination* (or *causal discrimination*). *Group fairness* and *individual fairness* represent their antonyms respectively. In both cases, fairness requires the specification of certain parameters of the model to study, called *discriminatory parameters*, with typical examples of 'gender', 'race', and 'age'. On one hand, for group fairness, the distribution of the software outputs for each of the groups, when grouping is done based on the values of the discriminatory parameters, should be similar. For example, suppose we set 'gender' to be the discriminatory parameter, ranging over two values of 'female' and 'male'. The LOAN software is said to be fair if it gives loans to the same fractions of both the applicant groups of 'female' and of 'male'. On the other hand, for individual fairness, the software should produce the same output for every two individuals who are identical except for the discriminatory parameters. For example, the LOAN software is said to be fair with respect to 'gender' if for all pairs of individuals, who differ only in the value of the 'gender' parameter but are identical for the other parameters, LOAN either gives the loan to all of them or none of them.

As this paper focuses on individual fairness, we formally state its concept. Let $P = \{p_1, p_2, \ldots, p_n\}$ be a set of *parameters* (or, *attributes*), for $n \in \mathbb{N}$, while we use p_i to indicate i-th parameter in P. Each parameter $p_i (\in P)$ is associated with a set of *values*, called the *domain* of p_i, denoted by $Dom(p_i)$, such that $(Dom(p_i))_{i \in n}$ is pairwise disjoint. The input space $\mathbb{I}$ of a set of parameters P is the Cartesian product of the domains of $p_1, p_2 \ldots p_n (\in P)$, i.e., $\mathbb{I} = Dom(p_1) \times Dom(p_2) \times \cdots \times Dom(p_n)$. An element I of $\mathbb{I}$ is a *data*, which we may also call *input* or *test case*. We use I_k as the value of the k-th parameter of input $I \in \mathbb{I}$. We also introduce $P_{disc} \subseteq P$ as the set of discriminatory parameters (e.g., gender, race, age, etc.). We interpret a ML model (*classifier*, for short), whose input space is $\mathbb{I}$, as function f; i.e., we use $f(I)$ to denote the result (i.e., decision) that the trained classifier f makes for input I.

Definition 1 (Discriminatory Input and fairness [27]) Let f be a classifier under test, γ be the pre-determined discrimination threshold (e.g. chosen by the user), and $I \in \mathbb{I}$. Assume $I' \in \mathbb{I}$ such that there exists a non-empty set $Q \subseteq P_{disc}$ and for all

$q \in Q$, $I_q \neq I'_q$ and for all $p \in P \setminus Q$, $I_p = I'_p$. If $|f(I) - f(I')| > \gamma$, then I is called a discriminatory input of the classifier f and is an instance that manifests the violation of (individual) fairness in f.

2.2 AEQUITAS *Approach*

AEQUITAS, developed by Udeshi et al. [27], is a black-box fairness testing technique to detect individual discrimination. AEQUITAS generates test cases by searching the input space of the software. The search process is separated into two steps, *global search* and *local search*, which are executed in this order. We review this search process in this subsection, with a focus on the global search. Global search is the target that our work tackles for improvement, whereas we use local search as it is without applying any modifications.

2.2.1 Global Search

Algorithm 1 describes the global search of AEQUITAS. The algorithm detects individual discrimination, inputting the set of all parameters (P), the set of discriminatory parameters (P_{disc}), and the number of test cases (N). After initializing the set of discriminatory inputs (*disc_inputs*) at line 3, the algorithm tries to detect discrimination using N test cases (line 4), as follows: The algorithm first selects one test case (I) randomly from the input space of the classifier (line 5). Then, it creates more input data from the test case (I), by extending it with all possible values of P_{disc}, and stores them in $I^{(d)}$ (line 7). Each of the created test cases $I' \in \mathbb{I}^{(d)}$ is run through the classifier, and the output is compared to the output of the execution of the original test case I. If the pair of outputs differs in more than the threshold value γ, the algorithm judges the test case I as discriminatory, and adds it to the discriminatory inputs list (lines 8 and 9). The algorithm executes this process for N test cases (i.e., for N trials).

Note that the global search is simply designed in the respect that the selection of N test cases are made randomly (at line 4). Note also that, since the algorithm takes the number of test cases N as an input argument, users can flexibly specify the number of test cases to control the search strength.

2.2.2 Local Search

After the global search, AEQUITAS tests the neighborhood of each input I from the discriminatory inputs found in global search by introducing a perturbation in the values of the non-discriminatory parameters in the input I. As stated in [26, 27], this local search is effective thanks to the robustness of machine learning classifiers, since a small variation in the input should not make a big change in the output.

In this case, if one input is discriminatory, by robustness, the inputs similar to that input are very likely to be discriminatory inputs as well. They propose three different approaches: *random*, *semi-directed* and *fully-directed*. Each approach differs in how the new inputs around I are generated. Experiments showed that the *fully-directed* approach outperforms the other approaches, hence we use the *fully-directed* approach throughout this paper.

Algorithm 1 Global Search [27]

procedure GLOBAL_EXP(P, P_{disc}, N)
  ▷ N is the number of trials in global search
  $disc\_inputs \leftarrow \varnothing$
  for i in $(0, N)$ **do**
    Select an input $I \in \mathbb{I}$ at random
    ▷ $\mathbb{I}^{(d)}$ extends I with all possible values of P_{disc}
    $\mathbb{I}^{(d)} \leftarrow \{I' | \forall p \in P \setminus P_{disc}.I_p = I'_p\}$
    if $\exists I, I' \in \mathbb{I}^{(d)}, |f(I) - f(I')| > \gamma$ **then**
      $disc\_inputs \leftarrow disc\_inputs \cup \{I\}$
    end if
  end for
  return $disc\_inputs$
end procedure

2.3 Combinatorial t-Way Testing

Combinatorial t-way testing (CT, or t-way testing) is a data sampling technique [18], based on t-way coverage criterion. CT assumes a test model (a.k.a., system under test (SUT) model), which is composed of two description components: (1) parameter-values and (2) constraints. The coverage criterion stipulates to cover all value combinations of size t of test model M.

Figure 1 shows an example of test model of CT [29], which specifies several computer configuration settings. The parameter-values of the test model specify that the test model has three parameters (CPU, OS, and BROWSER), which range over 2, 3, and 3 values, respectively. The constraints part specifies three constraints in Boolean formula. They represent forbidden combinations of the system, respectively capturing the following system specifications:

1. IE is available only for WINDOWS.
2. SAFARI is available only for MAC.
3. MAC does not support AMD CPUs.

Table 1 shows an example of a test set of test model of Fig. 1, satisfying the 2-way coverage criterion, which we call a 2-way test set. Each row of Table 1 represents a test case, which is a sequence of values (e.g., {AMD, WINDOWS, IE}) that is assigned

PARAMETER-VALUES

Parameter	Values
CPU	Intel, AMD
OS	Windows, Linux, Mac
Browser	IE, Firefox, Safari

CONSTRAINTS

1	(IE ⇒ Windows)
2	(Safari ⇒ Mac)
3	(Mac ⇒ ¬AMD)

Fig. 1 A test model example

Table 1 A 2-way test set of the test model in Fig. 1

No.	CPU	OS	Browser
1	AMD	Windows	IE
2	Intel	Windows	Firefox
3	Intel	Linux	Firefox
4	Intel	Windows	IE
5	Intel	Mac	Safari
6	AMD	Linux	Firefox
7	Intel	Mac	Firefox

to the parameters in the top column (i.e., {CPU, OS, Browser}). As it meets the 2-way coverage criterion, we can observe the test set covers all the value combinations of size 2 of the test model, that satisfy the given constraints, while it avoids to cover value combinations that violate the constraints. Note also that each test case in Table 1 satisfies the constraints.

As stated in [28], CT performs best overall when compared to other test generation techniques such as random testing or adaptive random testing.

3 Coverage-Guided Fairness Testing (CGFT)

Our goal is to improve the fairness-testing technique of Aequitas for finding discriminatory inputs. In order to improve it, we propose applying the concept of CT to Aequitas' global search, the first step in their search process, since global search only samples uniformly the input space. We call our proposed approach *Coverage-Guided Fairness Testing (CGFT)*. In this section, we explain the details of this approach. The three phases executed in CGFT are:

1. Obtain a test suite of specified size using CT
2. Find seed data by executing this test suite
3. Execute Aequitas' local search on these seed data

3.1 Obtain a Test Suite of Specified Size Using CT

3.1.1 Specifying Test Models of CT

Like the standard usage of CT, CGFT requires to specify a test model of the ML classifier under test. This model is defined by its parameter-values and a set of constraints.

Parameter-values
We derive the parameter-values of the classifier under test from its given training dataset, by inspecting the parameters and values. For example, the Census Income dataset [8] contains 14 parameters, including 'gender', 'race' and 'native country', which have respectively two, five and forty-one different values.

Constraints
These constraints are used to exclude some combinations of parameters when generating the test cases. Since CGFT uses AEQUITAS' local search, which perturbs the inputs by modifying their parameter values, these constraints may not be applied. Therefore, we do not consider constraints in this paper. However, we leave the inclusion of constraints to future work.

3.1.2 Combinatorial Test Generation of a Specified Size

After defining the model, we generate the test suite. One option is to generate the test cases solely based on a specified t value for t-way test suite generation. However, the number of test cases may be very rigid, e.g., as we shall show later in an example for the test model of Census dataset, 1-way may have 94 test cases, 2-way may have 5551 test cases, and 3-way may have 233,261 test cases. We believe that it would be better to have flexibility in the number of test cases generated, similar to Aequitas which lets users specify how many test cases to generate.

One obstacle of applying CT generation is its difficulty to control the number of test cases generated. CT tools, such as ACTS [17] and PICT [21], generate a test suite based on the information of the system and strength t, computing t-way test suite generation. In general, the standard usage of CT does not allow users to specify the number of test cases, since the number of test cases depends on the system itself. Thus it is not straightforward to obtain a test suite with a desired size. This inflexibility to control the number of test cases can be considered a disadvantage of using CT since, for example, in the global search of AEQUITAS users can specify the number of test cases (or inputs) N, flexibly controlling the testing process.

To overcome this obstacle in using CT, we implement an algorithm to control the number of test cases within CT generation regardless of the system under test. The basic idea is to allow the user to specify the number of test cases N, and obtain those N test cases from the CT-generated test suites. Our algorithm, called t-way+α, is based on the use of two different test suites, one of strength t and the other of strength

$t + 1$. This makes it possible to first cover all the test cases defined for strength t, and obtain the remaining test cases from the $(t + 1)$-way test suite. Thus, we ensure the testing of all the t-way combinations defined for the model.

The details of the algorithm are shown in Algorithm 2. For this algorithm, we added the following notations: $CT(t, M)$ stands for the test suite (obtained through CT) of strength t for the model M, *Alpha* is the test suite obtained by selecting N_α test cases randomly from the difference between the test suites of strength $t + 1$ and t (*Diff*), and TS_{res} is the test suite obtained after the execution of the algorithm, and it is a combination of the test suite of strength t ($CT(t, M)$) and the test suite *Alpha*. Our goal is to find the strength t whose test suite of strength $t + 1$ has more test cases than N. We start with $t = 1$ (line 3) and increase t until we find the t that satisfies that condition (lines 5 and 6). Then, we compute the difference between $t + 1$ and t test suites (*Diff*, line 8) and compute the number of test cases to be selected from this difference test suite (line 9). We obtain the test suite *Alpha* selecting N_α test cases randomly from *Diff* (line 10), and return the final test suite, of size N, combining $CT(t, M)$ and *Alpha* (line 12).

Algorithm 2 Combinatorial test generation: t-way+α

1: **procedure** CT_GEN(N, M)
2: $\quad TS_{res} \leftarrow \varnothing, Alpha \leftarrow \varnothing, Diff \leftarrow \varnothing$
3: $\quad t \leftarrow 1$
4: $\quad N_\alpha \leftarrow 0$
5: $\quad$ **while** $N > size(CT(t + 1, M))$ **do**
6: $\quad\quad t \leftarrow t + 1$
7: $\quad$ **end while**
8: $\quad Diff \leftarrow CT(t + 1, M) \setminus CT(t, M)$
9: $\quad N_\alpha \leftarrow N - size(CT(t, M))$
10: $\quad Alpha \leftarrow N_\alpha$ test cases from *Diff* randomly
11: $\quad TS_{res} \leftarrow CT(t, M) \cup Alpha$
12: $\quad$ **return** TS_{res}
13: **end procedure**

We exemplify the use of this algorithm using two execution trials of our evaluation from Sect. 4 for the test model of Census Income dataset. Suppose we set the number of test cases N to 3,000. Using CT, we generate 94 test cases for 1-way, 5,551 for 2-way and 233,261 for 3-way testing. Since $N = 3000$, Algorithm 2 uses both the 1-way and 2-way test suites. In particular, the algorithm takes all the 94 test cases from the 1-way test suite, and 2906 (this case's N_α) test cases from the 2-way test suite. Now, suppose that the number of test cases is $N = 7000$. Algorithm 2 uses both the 2-way and 3-way test suites, covering first all the test cases generated through CT with strength $t = 2$, and obtaining the rest from the 3-way test suite. More precisely, 5551 test cases are obtained from the 2-way test suite, whereas the remaining 1449 (this second case's N_α) test cases are obtained from the 3-way test suite. For both cases, the N_α test cases are first checked to avoid adding duplications (line 8).

3.2 Find Seed Data by Executing This Test Suite

Once we have a test suite of size N, we execute these test cases looking for individual discrimination, as described in Definition 1. We use the same algorithm as AEQUITAS' global search, Algorithm 1, but instead of selecting inputs randomly (line 5), we execute the test suite generated through CT, using Algorithm 2. If we find discriminatory inputs in this step, we call them *seed data*.

3.3 Execute AEQUITAS' Local Search on These Seed Data

Once we have obtained the seed data, we run AEQUITAS' local search on these seed data to find more discriminatory inputs. We use AEQUITAS' *fully-directed* approach since it outperforms the other AEQUITAS' approaches [27].

4 Evaluation

In this section, we evaluate CGFT to answer the following research questions (RQs):

RQ1 Can CGFT improve AEQUITAS in finding seed data?
RQ2 Can CGFT improve AEQUITAS in finding discriminatory inputs?
RQ3 How efficiently can CGFT work compared to AEQUITAS?

RQ1 explores the improvement introduced by CGFT when finding seed data, whereas RQ2, the main question in the evaluation, is focused on the impact this change in finding seed data has in the final result: the number of discriminatory inputs found. We need RQ2 since the seed data found by AEQUITAS may not necessarily be a subset of that found by CGFT. Finally, RQ3 is focused on the execution time of CGFT in comparison with AEQUITAS.

4.1 Experimental Setup

For a fair evaluation, the design of our experiment setup is based on the experiments executed in [27]. In our experiments, we conduct a fairness testing by using several ML classifiers (as used in [27]) and two datasets as training data (instead of one dataset used in [27]).

Datasets
The datasets used are Census Income dataset [8] and Statlog (German Credit Data) dataset [8], which are commonly used in fairness testing studies [1, 10, 27]. Census

dataset is used to predict whether a person's income exceeds $50,000 per year, and the size of this training dataset is around 32,000 samples. German Credit dataset is used to classify individuals as good or bad credit risks based on a set of properties, and the size of the dataset is 1000 samples.

Classifiers

Our experiments use the following five ML classifiers used in [27] from Python's Scikit-learn ML library [23], with its parameters defined in parentheses. The parameter values are the same as those used in [27], except for when they were not described in their work, for which we set the default values.

- SVM: (gamma = 0.0025)
- Decision Tree: (Default)
- MLPC: (solver = 'lbfgs', alpha = 1e−5, hidden_layer_sizes = (7, 5), random_state = 1)
- Random Forest: (Default)
- Ensemble Voting Classifier: (Random Forest, Decision Tree)

We excluded FairSVM [30] used in [27] from the experiments, since it is not in the Scikit-learn ML library. Moreover, Random Forest and Ensemble Voting classifier required an execution time longer than 10 h for each trial. Therefore, in our evaluation we present the results of the remaining three classifiers.

Discriminatory parameters

We set 'gender' as the discriminatory parameter for both Census and German Credit datasets. For German Credit dataset, we combined *male: divorced/separated, single* and *married/widowed* into one category, *male*, and merged *female: divorced/separated/married* and *single* into *female*, reducing from five to two gender values.

Conditions of experiments

We execute our experiments by executing the same number of test cases (N) in CGFT and AEQUITAS, allowing us to compare the impact CGFT has in finding discriminatory behavior using the same number of test cases.

Combinatorial testing

We obtain the CT test suites using the software ACTS [17].

Number of test cases

We chose the number of test cases N so that we can obtain two results for 1-way+α and two results for 2-way+α. For Census dataset, the 2-way test suite contains 5551 test cases; for German Credit dataset, it contains 6626 test cases. Thus, for both datasets, we conducted four experiments, where $N = 3000$ and $N = 5000$ use 1-way+α, and $N = 7000$ and $N = 9000$ use 2-way+α.

Machine specifications

All our experiments were performed on a machine with a 3.5 GHz Dual-Core Intel Core i7 processor having 16GB of RAM and running macOS Catalina 10.15.6. We implemented CGFT in Python.

4.2 Answers to Research Questions

Table 2 shows the results of the experiments. The rows represent the configuration settings, hierarchically structured by 'Datasets', 'Classifiers', and 'number of test cases (*N*)'. The columns show the number of seed data found, the total number of discriminatory inputs found, execution time and total number of inputs tested (test cases generated). The columns of 'AEQUITAS' and 'CGFT' contain the values obtained for each approach, with the Pct (%) column showing the ratio of CGFT over AEQUITAS. The column of 'Seed Data' for AEQUITAS represents the number of discriminatory inputs found during global search. The values in the table are the average of fifteen execution trials. For each comparison, the values in bold represent the approach with a higher value.

4.2.1 [RQ1] Can CGFT Improve AEQUITAS in Finding Seed Data?

From the columns of 'Seed data' in Table 2, we can observe that in 22 of 24 results (i.e., 91.6%), CGFT finds more seed data than AEQUITAS, with an average increase ratio of 198% and a maximum of 338% (for the case of Census dataset, MLPC and $N = 9000$). This means that, in 91.6% of the experiments, by introducing the strategy of CT in the global search of AEQUITAS, we increase the number of seed data found with same numbers of test cases N.

Answer: Yes, CGFT finds more seed data by an average of 198%.

4.2.2 [RQ2] Can CGFT Improve AEQUITAS in the Total Number of Discriminatory Inputs Found?

From the columns of 'Discriminatory inputs' in Table 2, we can also conclude the answer is affirmative. In this case, we can confirm that in all results (in all the prepared configuration settings), CGFT finds more discriminatory inputs than AEQUITAS, with an average improvement ratio of 130% and a maximum of 171% (for the case of Census dataset, MLPC and $N = 7000$). It is important to note that, even in the two experiments that CGFT found less seed data than AEQUITAS (German Credit dataset, SVM and $N = 3000$ and $N = 5000$), the total number of discriminatory inputs found is higher in CGFT than in AEQUITAS. This shows that, not only is the number of seed

Table 2 Results of the experiments

Dataset	Class	N	Seed data			Discriminatory inputs (total)			Time (s)			Inputs tested		
			AEQUITAS	CGFT	Pct (%)	AEQUITAS	CGFT	Pct (%)	AEQUITAS	CGFT	Pct (%)	AEQUITAS	CGFT	Pct (%)
Census	SVM	3000	6.0	**7.0**	117	231.4	**311.7**	135	**160.5**	142.5	89	12,745.8	**14,278.2**	112
		5000	9.1	**14.0**	153	279.5	**471.3**	169	228.4	**304.8**	133	19,275.4	**26,531.3**	138
		7000	12.6	**20.0**	159	368.3	**454.0**	123	354.0	**435.7**	123	26,482.2	**35,173.6**	133
		9000	16.3	**26.0**	159	419.1	**466.9**	111	412.3	**567.1**	138	32,782.9	**44,253.0**	135
	DT	3000	16.2	**50.0**	309	2094.4	**2630.7**	126	67.4	**171.9**	255	25,930.5	**55,817.8**	215
		5000	27.0	**77.0**	285	2221.2	**3364.6**	151	107.3	**276.2**	257	39,415.8	**66,721.7**	169
		7000	41.0	**112.0**	273	3184.6	**3830.1**	120	178.0	**415.5**	233	51,309.3	**89,417.7**	174
		9000	52.2	**141.0**	270	2998.8	**3497.6**	117	224.3	**531.6**	237	60,647.3	**98,320.5**	162
	MLPC	3000	5.9	**15.0**	253	265.7	**382.3**	144	35.7	**75.5**	211	12,437.5	**25,057.8**	201
		5000	9.3	**22.0**	236	314.5	**419.0**	133	57.4	**118.8**	207	19,351.5	**35,115.3**	181
		7000	13.9	**39.0**	280	359.7	**615.2**	171	83.2	**206.4**	248	27,364.9	**52,502.5**	192
		9000	17.7	**60.0**	338	444.9	**699.9**	157	106.5	**316.1**	297	33,918.2	**68,894.7**	203
German Credit	SVM	3000	**3.5**	3.0	85	108.5	**158.1**	146	**50.3**	30.3	60	**7386.5**	6763.5	92
		5000	**6.8**	6.0	88	180.9	**280.7**	155	**91.4**	63.1	69	**13,567.9**	12,700.5	94
		7000	7.4	**12.0**	162	200.5	**295.9**	148	106.9	**122.5**	115	16,611.1	**22,779.4**	137
		9000	10.1	**18.0**	178	271.4	**306.5**	113	145.0	**179.8**	124	22,065.3	**31,880.1**	144
	DT	3000	28.9	**63.0**	218	2291.5	**2367.5**	103	244.2	**410.2**	168	33,483.8	**55,262.8**	165
		5000	45.7	**110.0**	241	2193.3	**2547.2**	116	308.1	**700.9**	227	46,742.3	**63,994.0**	137
		7000	62.8	**148.0**	236	2295.9	**2602.3**	113	436.5	**932.4**	214	57,440.3	**65,242.1**	114
		9000	86.7	**198.0**	228	2439.9	**2639.4**	108	588.8	**1293.7**	220	**70,939.7**	70,081.1	99
	MLPC	3000	247.0	**304.0**	123	1569.8	**1769.1**	113	2317.6	**2680.4**	116	**16,495.9**	14,820.3	90
		5000	404.9	**504.0**	124	1685.6	**1972.7**	117	3806.5	**4396.9**	116	**18,633.6**	17,069.7	92
		7000	575.5	**704.0**	122	1876.2	**2209.7**	118	5373.9	**6384.7**	119	**21,425.1**	18,718.1	87
		9000	722.6	**888.0**	123	2076.8	**2359.8**	114	6924.4	**8081.4**	117	**24,116.4**	20,984.3	87
Average					198			130			171			140

data important when testing, but also the seed data itself. Therefore, even though in some cases we find even less seed data, CGFT is able to detect more discrimination after executing local search.

Answer: Yes, CGFT finds more discriminatory inputs in total by an average of 130%.

4.2.3 [RQ3] How Efficiently Can CGFT Test Compared to Previous Work?

The columns of 'Time' in Table 2 show that the execution time of CGFT is higher than AEQUITAS' in 21 out of 24 experiments. The execution time of both CGFT and AEQUITAS depends heavily on the total number of inputs generated ('Inputs Tested' columns in Table 2): as a general rule, if more test cases (and hence, inputs) are generated, then the execution time will be longer. The exceptions are Census dataset's SVM $N = 3000$, where CGFT is faster even though it generates more inputs, and German Credit dataset's DT $N = 9000$ and all MLPCs, where CGFT takes more time even though it generates less inputs.

This fact could be considered a downside of CGFT. However, in all the execution trials, most of the execution time was spent in local search. We have not worked on this phase yet, since we made use of AEQUITAS *fully-directed* local search, therefore we will leave any improvement in time performance as future work, as well as creating other experiments to clearly assess the efficiency of each approach.

Answer: CGFT usually requires more execution time than AEQUITAS by 170% in average. However, most of the execution time is spent in local search, which is the same as AEQUITAS' implementation.

5 Threats to Validity

Datasets
Our experiments used two different datasets. The conclusions obtained from our experiments thus may not be generalized. However, in most fairness testing works, one or two datasets are used in the experiments [10, 27]. We leave further experiments with more datasets to our future work.

Classifiers
We tested five different classifiers in [27] for each dataset in our evaluation experiments. However, two of them required an execution time of more than 10 h, exceeding our timeout. Also we excluded Fair SVM [30], the other classifier used in [27], since it is not included in Python's Scikit-learn ML library. Our evaluation is thus more limited than that of [27] in terms of number of classifiers. To complement this limita-

tion, we used two different datasets, which diversify our experiments with 6 different configurations (while AEQUITAS is evaluated with only Census dataset [27]).

Discriminatory parameters
As of now, we only tested CGFT with datasets in cases where we selected *gender* as the discriminatory parameter. Even though CGFT's implementation is generic and can be used with any discriminatory parameter, we will consider testing other discriminatory parameters in the future, as well as including the case of having more than one discriminatory parameter.

Construct validity
If our CGFT implementation has bugs, the results would be different. We cannot completely rule this out, but we have tried our best to assure that their are no bugs in the CGFT implementation.

6 Related Work

Studies on software fairness testing mainly focused on group fairness and discriminations, before the concept of individual one was proposed [5, 14, 15, 31]. For example, the technique proposed by Calders and Verwer [5] makes data groups, which are grouped by the values of the target discriminatory parameter, and check if the fractions of the software outputs for each of the groups should be similar. To do so, they introduce a metric, called the Calders-Verwer (CV) score, which shows the difference between the largest and smallest outcome fractions.

The concept of individual fairness (a.k.a., causal fairness) was introduced by Galhotra et al. [10], in order to address limitations of group fairness. The advantages of the individual fairness include that it can detect some discriminations that the group fairness cannot. They also developed a testing method, named THEMIS, which includes a test generation algorithm dedicated to detect individual discriminations. The test generation algorithm, however, is based on random sampling, similar to the global search of AEQUITAS.

Udeshi et al. [27], where AEQUITAS is developed, is based on [10] and its developed tool THEMIS. AEQUITAS provides a directed algorithm for test generation for efficiently detecting more individual discriminations. Moreover, they propose a mechanism to improve fairness of the classifier using the detected discriminations to retrain it. Our work aims to improve test-generation mechanism part of AEQUITAS using combinatorial t-way testing. Since our test-generation technique (i.e., CGFT) can find more discriminations, we think it may also improve the retraining mechanism of AEQUITAS. We leave investigating this conjecture to our future work.

Aggarwal et al. [1] propose a black-box fairness testing technique targeting the individual discrimination. Their approach is also divided into two processes, global search and local search. The combination of symbolic execution technique and an explainability technique in the global search phase, while in the local search phase they toggle the non-protected attributes to find more discriminatory inputs. This approach differs from CGFT in how the test suite is obtained, since CGFT uses CT as the main technique. We did not evaluate CGFT against the fairness testing technique of Aggarwal et al., although they provide a similar functionality of detecting individual discriminations. This is mainly because that the source code of technique by Aggarwal et al. seems not available, and thus it is difficult to compare them with our evaluation based on experiments.

Fliptest [4] obtains a test suite from a real life database, such as the training dataset, and creates counterfactual tests based on observational data, using optimal transport. Their optimal transport algorithm takes two different distributions, e.g. male and female, and matches each individual in each distribution to the equivalent individual in the other distribution. Their proposal uses real life data, whereas we generate the test cases.

Removing bias is another branch of software fairness, and it aims to build a fairer model. These algorithms can be classified into pre-processing, if the data is transformed before training the classifiers e.g. [2, 14]; in-processing, if the training algorithm is modified to reduce the bias by, for example, imposing certain constraints or modifying the objective function e.g. [3, 16]; and post-processing, by reassigning the labels assigned by the classifier e.g. [12, 24]. Moreover, other techniques, such as Fairway [6], mitigate bias through several strategies, combining pre-processing and in-processing methods.

As for combinatorial testing, Ma et al. use combinatorial t-way testing for testing DNN [19]. Their focus, however, differs from ours in several aspects, including that it focuses on using combinatorial t-way testing to cover neurons of DNN, not on fairness testing.

7 Conclusions and Future Work

In this paper, we proposed a black-box fairness testing technique for machine learning models, named Coverage-Guided Fairness Testing (CGFT), that incorporates the concept of combinatorial t-way testing [18]. CGFT is designed based on AEQUITAS [27], a state-of-the-art black-box fairness testing technique. Given that the test-generation mechanism of AEQUITAS is divided into two phases of global and local search, we apply combinatorial t-way testing to the global search, since AEQUITAS simply uses random sampling. In developing CGFT, we also devise a technique that can flexibly control the number of test cases generated within t-way test generation, incorporating CT in AEQUITAS. We evaluated CGFT by experiments, where we used two datasets and several classifiers with each dataset. The experimental results confirmed that CGFT outperforms AEQUITAS in detecting more discriminatory inputs by an average of 130%, at a reasonable computational overhead.

For future work, we plan to evaluate CGFT by using more datasets and classifiers, to further generalize the evaluation. We also plan to evaluate CGFT in comparison with another state-of-the-art black-box fairness testing technique by Aggarwal et al. [1], which uses dynamic symbolic execution and local explainability, for their discrimination detecting capabilities and computational costs. Another direction of further evaluation is to evaluate the retraining process of AEQUITAS and whether it can be improved by using CGFT, since the work in this paper focused on its discrimination-detecting capability. For the technical advancement side of CGFT, we are interested in making use of constraints in CT, as we make no use of them on this paper. We expect that an effective use of constraints in t-way test generation can alleviate generating un-realistic data that contains impossible combinations, and thus might improve the test generation and retraining process. We are also interested in developing a technique to combine the use of CT of CGFT and the use of dynamic symbolic execution of [1] in the global search phase of AEQUITAS. Moreover, in this paper we only addressed the global search process in AEQUITAS, thus we leave any improvement in local search as future work.

References

1. Aggarwal, A., Lohia, P., Nagar, S., Dey, K., Saha, D.: Black box fairness testing of machine learning models. In: Proceedings of the ACM Joint Meeting on European Software Engineering Conference and Symposium on the Foundations of Software Engineering, ESEC/SIGSOFT FSE , 2019, pp. 625–635. ACM (2019). https://doi.org/10.1145/3338906.3338937
2. Bellamy, R.K., Dey, K., Hind, M., Hoffman, S.C., Houde, S., Kannan, K., Lohia, P., Martino, J., Mehta, S., Mojsilovic, A., et al.: AI fairness 360: an extensible toolkit for detecting, understanding, and mitigating unwanted algorithmic bias (2018). arXiv preprint arXiv:1810.01943
3. Berk, R., Heidari, H., Jabbari, S., Joseph, M., Kearns, M., Morgenstern, J., Neel, S., Roth, A.: A convex framework for fair regression (2017). arXiv preprint arXiv:1706.02409
4. Black, E., Yeom, S., Fredrikson, M.: Fliptest: fairness testing via optimal transport. In: Proceedings of the 2020 Conference on Fairness, Accountability, and Transparency, pp. 111–121 (2020)
5. Calders, T., Verwer, S.: Three Naive Bayes approaches for discrimination-free classification. Data Min. Knowl. Discov. **21**(2), 277–292 (2010)
6. Chakraborty, J., Majumder, S., Yu, Z., Menzies, T.: Fairway: a way to build fair ML software. In: Devanbu, P., Cohen, M.B., Zimmermann, T. (eds.) ESEC/FSE'20: 28th ACM Joint European Software Engineering Conference and Symposium on the Foundations of Software Engineering, 2020, pp. 654–665. ACM (2020). https://doi.org/10.1145/3368089.3409697
7. Dastin, J.: Amazon scraps secret AI recruiting tool that showed bias against women (2018). https://www.reuters.com/article/us-amazon-com-jobs-automation-insight-idUSKCN1MK08G. Accessed: 24 Feb 2021
8. Dua, D., Graff, C.: UCI machine learning repository (2017). http://archive.ics.uci.edu/ml
9. Ferral, K.: Wisconsin supreme court allows state to continue using computer program to assist in sentencing (2016). https://madison.com/ct/news/local/govt-and-politics/wisconsin-supreme-court-allows-state-to-continue-using-computer-program/article_7eb67874-bf40-59e3-b62a-923d1626fa0f.html. Accessed: 31 Mar 2021
10. Galhotra, S., Brun, Y., Meliou, A.: Fairness testing: testing software for discrimination. In: Proceedings of the 2017 11th Joint Meeting on Foundations of Software Engineering, pp. 498–510 (2017)
11. Goodall, N.J.: Can you program ethics into a self-driving car? IEEE Spectr. **53**(6), 28–58 (2016)
12. Hardt, M., Price, E., Srebro, N.: Equality of opportunity in supervised learning (2016). arXiv preprint arXiv:1610.02413